T0195906

BIG BANG PROBLEMS

HOW, WHEN, AND WHERE
THE UNIVERSE BEGAN

BIG BANG PROBLEMS

HOW, WHEN, AND WHERE
THE UNIVERSE BEGAN

Bobby McGehee

authorHOUSE®

AuthorHouse™
1663 Liberty Drive
Bloomington, IN 47403
www.authorhouse.com
Phone: 1-800-839-8640

Published by AuthorHouse   12/15/2015

ISBN: 978-1-4969-4646-1 (sc)
ISBN: 978-1-4969-4641-6 (hc)
ISBN: 978-1-4969-4642-3 (e)

Print information available on the last page.

Supporting books by Bobby McGehee. Available at most book outlets; all are recommended for reference. All are also available from the publisher in paperback, hardback, and e-book, at: <www.authorhouse.com/bookstore>

Three related books and their ISBN:

1. **New Universe Theory**. (N.U.T.) Written during the development of this new theory of the universe. Theory. Displaces the 1920's big bang idea in favor of reality. [©2000]
Softcover: 9781418494292
Hardcover: 9781418494308
Ebook: 9781418494315

2. **Model of the Universe**. Conveys geometry of the universe. Presents proof (40 year time span data) the universe is not expanding with acceleration. Uses Palomar data. [©2005]
Softcover: 9781449067922
Hardcover: 9781449067939
Ebook: 9781449067946

3. **Big Bang Problems / How the Universe Began**. Presents several big bang credibility problems and proves the universe's growth is not accelerating, It is decelerating proven over a 50 year time span using new (Spitzer) data. Book presents recent discovery about the universe origin site(s?) with supportive analyses. [©2015]
Hardcover: 9781496946416
Softcover : 9781496946461
Ebook: 9781496946423

Contents

Figure Index: Tables, Graphs, and Sketches

Bobby McGehee

Preface

Scientific thinking uses logic with only proven facts and verified phenomena, which are laws of physics. This new concept is consistent with the laws of physics, and the big bang theory is not. Several problems and proofs are presented that disprove the big bang theory and support this consistent new concept in the law of physics.

Many scientific thinkers have never been comfortable with the big bang origin concept. In deference to earlier cosmologists, who, without a scientific theory available that is totally consistent with laws of physics, it is now understandable how a nonprovable theory was accepted as an explanation. Now, we have a provable concept that is consistent with proven laws of physics. It requires only scientifically valid interpretation of observations. No magic or myth. Before the 1930s, when the big bang theory was conceived, the energy and matter source from annihilating positroniums were not yet discovered. As stated, arguments are presented that disprove the big bang theory, but the latest undeniable proof is presented in Argument/Problem #8. The data analyses use credible six-digit resolution redshift data that was acquired with Spitzer Space Telescope during the first decade of 2000. The new data is compared to fifty years ago (1959 Palomar data) and proves the objects (galactic clusters) are decelerating, not accelerating from or toward mythical dark energy.

Now, the origin site void discovered in 2007 fits the new theory nicely. The site is located in the same direction (Problem #5) as predicted based on analyses of 1959 Palomar Telescope data cataloged by Abell, Corwin, and Olowin in 1959 and measured again by Struble and Rood in 1999, which, when compared, shows no acceleration.

The Spitzer Space Telescope[1] six-digit higher resolution redshift data (2009) allows analyses that shows the galactic clusters are decelerating, as this new concept predicts. These analyses strongly support, if not prove, the new concept.

---

[1]   Spitzer does not orbit the Earth. It orbits the sun in the Earth's shadow.

Acknowledgments

This book exists thanks to help from many people. Many contributed to our knowledge, and several have never been appropriately credited. Yet some have taken credit for others' accomplishments. Many have commented that they could never accept the big bang theory, as it is not plausible because it is not possible. Thanks to all. The following is a partial list of important credits to this book.

First, my wife Nancy has given me her relentless support. We met in English composition class at Oklahoma State University in 1950. When you find grammatical errors, it is because I was more interested in flirting with Nancy than in the teacher's lectures. She now is a recovering Alzheimer's patient, and I am honored to be her full-time caregiver.

Spitzer Space Telescope staff and archive teams makes it possible with six-digit resolution redshift to prove the objects in the universe are decelerating apart, not accelerating. Analyses results are presented in this text, along with verifiable via data in publically accessible data catalogues. Spitzer deserves NASA's full financial support.

Robert E. Farrell, doctor of engineering, collaborated on the McGehee concept after publication of the second book on this subject. He assisted in calculations and graphs for confirming and determining rates of decelerations of galactic clusters. He makes presentations of this new theory at conferences and seminars. He contributed support throughout with analyses and documentation.

Edward Ryder, retired USAF colonel; former White House military advisor; past vice commander of NATO F-16 base, Aviano Air Base in Italy; and current manager training foreign buyers of F-16 and F-35 aircraft for Lockheed Corp. Ed provided peer review and helpful suggestions.

Harold G. Corwin Jr., doctor of physics and astronomy from Edinburg, joined the Abell, Corwin, and Olowin (ACO) team to compile data from various sources, making it possible for others to accomplish studies. Sources include data from the Palomar All Sky Survey (PASS). That was the first application for which the then world's largest telescope was dedicated. I had made inquiries to support my studies, and he was the only team member I was able to contact. Abell passed away in 1983, and Olowin is from the University of Oklahoma. Corwin tells me that Olowin is now at St. Mary's College in California. Harold Corwin gave me the websites that allowed me to achieve the studies presented in *Model of the Universe*. He retired in 2012.

I contacted John Huchra, doctor of astronomy and physics at Harvard University, in 2005. I asked for information to help with my studies, and he wrote me before and telephoned after he had a stroke, indicating he was being less active. He has been the coordinator for several astronomy research teams. I have not been able to reestablish contact, and now to my dismay, I learn that he died in 2010, only a few months after he told me he had also tried to prove that galactic clusters are not accelerating apart. (Harold Corwin corrected my memory that it was not himself that had tried to analyze galactic cluster data to show no accelerating separation.) John Huchra would have been pleased with the proof presented in this document that galactic clusters are not accelerating apart.

William G. Ryder, nuclear engineer, assisted in calculations of galactic cluster velocities to verify there is no outward acceleration of all two-hundred-forty-six common[2] ACO cataloged galaxies over a forty-year period. Twenty-four decimal place data calculations were needed. We both accomplished these together on separate computers and kept repeating the calculations until we both could produce identical results to the detail needed. Results are presented in *Model of the Universe* that show no acceleration, as Hubble claimed.

James C. Baker, Boeing and NASA rocket scientist/engineer, made theory suggestions. He made the credible statement, "It makes more logical sense to assume matter pre-existed the universe than to assume all mass and energy came from a single point."

Les Sherry, industrial engineer and California State University at Fresno (1962) and retiree (1995) manager of the US Navy shipyards in California, is the previous president of Arizona West Valley Engineers Organization. Les peer-reviewed this document and provided several helpful suggestions.

I give credit and thanks to the unknown artist who produced the rendition of the void picture used in this document. This is a good interpretation of what I think the way the origin void would appear. The stars at the edge of the void would be much smaller than as shown. This picture appears in several sites on the Internet, but none identify the producer. Lawrence Rudnick informed me that the NASA/NRAO organization produced the rendition.

---

[2]   This is common in both the pre-1959 Palomar original ACO galactic cluster catalog and the 1999 S&R Palomar survey

Hypothesis

Most scientific thinkers have never been satisfied with the big bang theory, which is in direct conflict with many laws of physics. This new concept was developed from facts and observations, and it is provable. We assume the universe is a spherical volume within an endless array of primordial matter that are equally spaced positroniums. These smallest of all atoms are mutually orbiting positrons and electrons (matter and antimatter), spaced such that it calculates to the same mass density as the universe today. These smallest of atoms are stabilized by centrifugal forces while held together by their mutual gravitation and electrostatic charges. Stability is by synchronous rotations throughout primordial space. Outside the universe in primordial space, there are no extraneous radiations to upset the balance; therefore, they are stable. Yet an initial annihilation in primordial space upset the balance and triggered a cascading wave of annihilations propagating in all directions at near the speed of light. A void was left at the site.

Forward-radiating annihilation photons overlap and precipitate into subatomic particles, which coalesce by velocity-enhanced gravitational attractions. As particles accrete, they decelerate by the transfer of linear momentums into angular momentums. Objects grow/evolve into astronomical-sized objects. Rearward-radiating annihilation photons do not adequately overlap and are energy packets to heat dark matter, which radiate microwave background radiation.

Cosmology: Scientific Study of the Universe

For thousands of years, humankind has gazed up into the heavens at the beautiful Milky Way galaxy and pondered "From where did it come?" Over the years, as more knowledge was acquired, our views of the universe have changed. Our thinking has evolved from a belief that the Earth was flat, and then it was thought the Earth was the center of the universe. Then the misinterpretation of redshift gave the appearance that everything erupted from a single point. Thinking evolved to the idea of a big bang belief that everywhere is the center of the universe and it is expanding without limit. Now again, we have reached dogma. Scientific thinkers recognize the big bang theory has serious credibility problems.

It is now proved and recognized that the big bang concept was inferred from misinterpretation of redshift observations first acquired in the early 1910s at Lowell Observatory in Flagstaff, Arizona. The big bang theory was concocted before positron/electron combination 'positroniums' were known. These smallest of atoms were discovered in the 1930s and not understood until the 1950s. Positroniums have a tremendous amount of energy, and when their positrons and electrons come into contact, they annihilate, the conversion of all their mass into energy that is evolving into all the mass and energy in the universe.

The initial annihilation started the cascading at what is now the center of the universe. It is a huge void in the Eridanus constellation, about 10 billion light-years from Earth. Now, we have a provable scientific explanation. Finally, it all makes sense without myth.

Definitions

Asymmetric: Not balanced or regularly arranged on opposite sides of a line or around a central point

Belief and faith: Accept without proof

Deflagration wave: A universe-encompassing shell propagating and cascading outward at the speed of light; the transcending process where primordial matter converts into other corporeal (real) matter; both mass and energy[3]

Laws of physics: Principles verified many times and never disproved

Matter: Includes all mass and energy; all particles and material as well as all forms of energy; can be converted from one form to another but cannot be created nor destroyed

Myths: Ideas that cannot be proved or disproved

Primordial universe: An infinitely large volume that includes all primordial matter surrounding the universe within; includes all time, mass, and energy in all directions from minus infinity to plus infinity

Scientific thinking: Mental process that uses logic with only proven facts and phenomena

Universe: A region within the primordial universe that encompasses all energy, mass, space, and time

---

[3]   Physical processes are described in part II.

Laws of Physics Pertinent to Universe Origin

1. Motion (Isaac Newton) $F = ma, = d(mv)/dt$
   - Force is required for acceleration
2. Energy and mass (Albert Einstein) $E = mc^2$
3. Force of gravity (Isaac Newton) $F_g = m_1 \times m_2/d^2$
   - Mutual attraction all objects
4. Relativity (Albert Einstein, 1905) $m = m_0 / (1 - m^2/c^2)^{1/2}$
   - Velocity effect on mass and time
5. Continuity (Rudolf Clausius, 1850) $M_1 = M_2$
   - Mass and energy into a system = mass and energy out + mass and energy remaining in the system
   - First law of thermodynamics
   - Succinctly, matter cannot be created or destroyed
6. Schwarzschild radius $R_S = 2Gm / c^2$
   - Radius at escape velocity ($\leq c$)
   - Escape velocity $V_{esc} = (2Gm / r)^{1/2}$
   - Velocity required for two objects to escape from their mutual gravitational attraction
   - For a black hole, $V_{esc} = c$ (impossible to achieve)

Nomenclature

a: acceleration ($\sim dv/dt$)

c: velocity of light

e: entropy

E: energy

F: force

G: universal force of gravity = 6.67 x 10^{-11} ; Nm^2 / kg^2

m: mass

M: matter; mass+ energy

N: Newton unit of force (Kg meters / sec^2)

r: radius

Rs: Schwarzschild radius

v: velocity (speed)

V: velocity (vector)

Part I: Big Bang Credibility Problems

Introduction

The big bang theory was conceived based on misinterpretation of redshift data that was initially acquired in the early 1910s by V. M. Slipher of Lowell Observatory at Flagstaff, Arizona. His data correctly revealed "the farther away a galaxy, the faster it is receding." That superficially implied the universe is expanding. In the 1920s, other astronomers further studied the phenomena, and they misinterpreted the data to indicate the universe is accelerating outward. Running their clock backward led to their erroneous conclusion that all matter, mass and energy, in the universe began at a single point 13.7 billion years ago.

The best-known outspoken dissenter was Sir Fred Hoyle, who derogatorily said the concept is just a bunch of "big bang" bunk. The name stuck. Many scientific thinkers have never accepted the accelerating expansion theory because it is inconsistent with laws of physics and requires mythical phenomena. All astronomers recognize that distant galaxies are indeed receding at rates that are higher at greater distances. But the farther away, higher velocity clusters are younger. When specific galactic clusters are observed over time, redshift shows they have slowed and are slowing.

Now, deceleration is proven when comparing the same galaxy cluster using fifty-year time-lapse data from the same specific ACO galaxy clusters.[4]

---

[4]   This is named for the team who originally identified and cataloged the clusters from an important part of Palomar All Sky Survey (PASS): G. O. Abell, H. G. Corwin, and R. P. Olowin

Comparing specific verified original 1950s ACO data with recent high-resolution redshift data disproves acceleration. This text reveals the results of these analyses. The first ACO cluster redshift data were acquired in 1959 from the PASS and observed again in 1999 by Struble and Rood. Their redshifts were again measured with the high-resolution Spitzer Space Telescope between 2003 and 2009 (cataloged in 2009).

Measurements of ten of the largest ACO galactic clusters include about twelve hundred galaxies (200 billion to 400 billion stars each) reveals a fifty-year average redshift decrease of 0.0001694, a velocity decrease of 0.002138 Km/second at 0.3 redshift. We now know the universe is not accelerating. Conceived before the availability and awareness of these facts, the Hubble big bang equation presumptuously predicts a forty-year velocity increase of 0.000245 Km/second.

Deceleration begins when speed of light photon matter first precipitates into mass matter, for example, neutrinos, bosons, fermions, and other subatomic mass particles. The gravities (velocity-enhanced) of all these particles cause coalescence, accretions, merging, collisions, and mutual orbiting. All of these physical processes extract linear momentum and convert it into angular in accordance with the laws of physics.

The universe continues to grow, as Slipher's original (1912) redshift discovery indicates. That observation—plus these analyses—conclusively contradicts and disproves the big bang theory, dark energy, and inflation ideas.

As previously stated, the big bang theory was well intended but based on misinterpretation of data and includes many laws of physics violations. The big bang provided astronomers with a stop-gap explanation until Paul Dirac and others discovered positrons and positroniums in the early 1930s. Problems/proofs are listed in part I—Problems #1–10—which justify superseding the big bang theory with a better and feasible scientific concept.

Recent high-resolution Spitzer Space Telescope data, measured between 2003 and 2009, compared to verified Palomar 1959 and 1999 data, are presented and provide conclusive proof that the big bang theory is not valid.

Problem #1: Continuity

The first law of thermodynamics ($M_1 = M_2$) succinctly states that matter (mass and energy) cannot be created or destroyed. The thermodynamics textbooks state it in full as, "All matter (mass and energy) entering into a system is equal to the mass and energy leaving the system plus the mass and energy that remains in the system."

According to the second law of thermodynamics, entropy is forever increasing ($E_1 > E_2$). Energy, a form of matter, is proportionately decreasing. The big bang concept contrarily says the universe is expanding and gaining more momentum, a form of energy with time, from the invented and fictions termed as "dark energy."

Redshifts of galactic clusters at various distances show galactic clusters at greater distances are receding at higher speeds. The apparent but erroneous interpretation concludes that all objects are accelerating outwards. When hypothetically running their clock backward, it appeared everything is emanating from a single point 13.77 billion years ago. They called that the "big bang."

However, when we measure again the same objects after several years with six-digit redshift resolution, we find (1) their velocity outward has not increased or (2) their speed/redshift outward has decreased. Simply put, the universe's assumed acceleration is imaginary, and the assumption was intended to explain the misinterpretation of observations by early twentieth-century astronomers. The big bang assumption does not suggest a viable source for matter. The term "dark energy" was invented to explain an apparent phenomena that does not—and by the laws of physics—cannot exist.

Problem #2: Ignoring Speed of Light as a Limit for Expansion

At the speed of light, any physical object's mass becomes infinite. Albert Einstein originally wrote the relativity equation, and it is as follows: $[M = M_0/(1 - v^2/c^2)^{1/2}]$

Overlooking all of the other impossibilities for the big bang to be real, it would require considerably more time than allowed by the big bang's maximum age for the universe to evolve into what we observe. Based on the big bang theory, all mass and energy erupted from a point source (unity). Backtracking along the projected illusion of universe expansion, the universe's age appears to be 13.77 billion years. To avoid facing up to the big bang being impossible by time constraints, there was an idea of inflation invented. It is said to have evolved/progressed faster than the speed of light. The fact that such speeds are an impossibility is explicitly shown by Einstein's relativity equation, which has since been proven many times at centrifuges and other laboratories around the world.

NASA's Chart Summary (The Chart Authors' Explanation)
They claim that, after the near-infinite temperature big bang theory matter supposedly cooled to a temperature of greater than 10^{32} degrees Kelvin, it cooled to approximately 10^{24} degrees Kelvin. The energy then transmuted into sub elementary particles (for example, bosons, quarks, mesons, and so forth). As it cooled to about 1012 degrees Kelvin, sub elementary particles fused into leptons as it cooled to 10^9 degrees Kelvin. As matter mixed and cooled toward approximately greater than 3×10^3 degrees Kelvin, leptons fused into hadrons and then combined into atoms. After matter further cooled to less than 10^3 degrees Kelvin, clumping began.

Bobby McGehee

Figure I-1. NASA's history of the universe
chart. Taxpayer-funded by DOE and NSF.

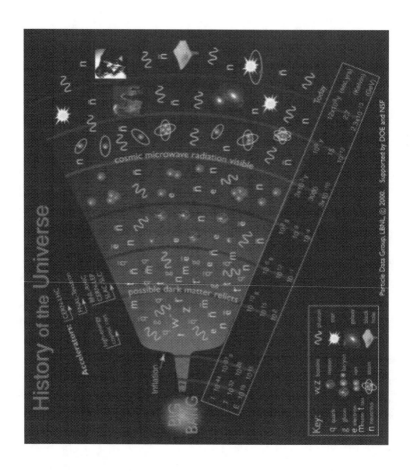

This chart artistically illustrates Steven Weinberg's concept (reference 5) of matter transmutations, but the parts are not flying apart and accelerating, as shown in this/NASA's chart.

After the postulated big bang theory's presumed near infinite energy source (impossible) converted a single point into all of the mass objects and energy (preposterous) in the universe of today, it also clearly points out a so-called inflation period, the time when

12

they say the universe expanded faster than the speed of light. All creditable physicists know that such expansion is impossible by the proven laws of physics.

Weinberg's concept of energy-to-mass uses gravity-enhanced fusions, but they say particles must be decelerated via coalescence and mutual orbiting before fusions. The big bang concept requires particles to be expanding (accelerating) apart that inhibits, if not precludes, coalescing into larger particles, like in collider centrifuges where particles are fractured and fragmented. Not fused at the big bang theorized extreme are thermal-produced, divergent, high-particle velocities.

Bobby McGehee

Problem #3: Schwarzschild Radius

Schwarzschild radius is the radial distance from the center of gravity of any mass object to where their escape velocity, from their mutual gravitational attraction, exceeds the speed of light (impossible). If any body of mass is concentrated to/within its Schwarzschild radius, its outside surface becomes a one-way barrier, and nothing can ever escape again, including light. It is known thereafter as a black hole. Many such objects are known to exist in the universe, and more continue to be discovered and observed.

To accommodate the big bang concept, these proven phenomena are ignored and violated for all of the universe's mass and energy. Such an event would require impossible higher-than-light speeds for any of the universe's matter to escape from its original theorized unity (from matter's self gravity). The mass would become infinite for any object achieving the speed of light, first expressed by Einstein's relativity equations (Problem #2). It has been proven many times.

The Schwarzschild radius for the solar system calculates to be about the size of a basketball. The mass observed in the universe is many trillions of times greater than the solar system. According to the big bang theory, all of the matter in the whole universe was once within a much smaller dimensionless dot, often referred to as "singularity." The proven fact is, if it were so compacted, it could never have escaped its initial or physical location and size.

Again, by the proven relativity equations, stated many times, not any mass's velocity (v) can ever equal the speed of light (c). Therefore, [v < c], always. And in conclusion, if the universe started as unity, all the universe's mass would still be inside its original unity source.

Problem #4: Newton's Laws of Motion

Force and energy are required for velocity change, acceleration or deceleration (F = ma and/or F = d(mv)/dt).

The big bang origin concept requires changes in velocity from zero to higher-than-light speed. Then all matter slowed. Now, the big bang theory claims the universe's contents are again accelerating. As recent as 2011, some have claimed to prove the existence of dark energy to produce renewed and continued expansion, even at extreme redshifts.

Big bang theory advocates claim, using some recent (even since 2000 AD) observations, and support the idea that matter in the universe is accelerating apart. Preposterous! Isaac Newton defined the first scientifically proven law of physics/laws of motion in the 1600s. Like all other laws of physics, these have been proven many times and never been disproven, (F = ma) and (F = d(mv)/dt). The misinterpretation of measured "change in redshift with distance" was construed to also mean individual galaxy's increase in velocity with time. Many astronomers accepted that misinterpretation, and the same erroneous interpretation is being made again. We now have the data that proves that was simply misinterpretation and the acceleration conclusion is not valid. (Figure II-1 illustrates an analogy, which shows how that misinterpretation occurred.)

A dark energy team of astronomers was established to investigate that hypothetical phenomena, all because of what appears to be accelerating expansion of the universe. A prize was recently awarded for their dark energy studies. No one should insult those who have put forth much effort to explain the mythical phenomena, dark energy. Hubble's earlier misinterpretation led them astray, and they are now victims of dogma.

Just because we observe an object that is traveling in the same direction and faster than closer objects, that does not prove the closer/ slower objects will accelerate and increase their speed and defy Newton's laws of motion.

The American Association for Advancement of Science (AAAS) publication, *Science* (August 6, 2004), questions the validity of the proof of dark energy, even though several reputable astronomers supported the contrary observations. The apparent new evidence of accelerating expansion was observed using new powerful instruments, acquiring data at the farthest reaches of the observable universe. Even though the instruments used are of the utmost accuracy, all of the evidence is nothing more than another misinterpretation of data, just like Hubble was also a victim in the 1920s by misusing valid data that Lowell Observatory's V. M. Slipher originally observed in the 1910s.

Problem #5: Cosmic Microwave Background (CMB) Radiation

Observation of cosmic microwave background (CMB) proves there is background radiation, but it does not prove its source. No one denies the existence of the background radiation, nor the measured spectral distribution and how well it matches the Plank black body radiation equation.

<u>Figure I-2</u>. CMB. Cosmic Microwave Background

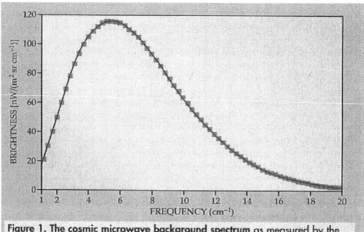

Figure 1. The cosmic microwave background spectrum as measured by the *Cosmic Background Explorer's* FIRAS spectrometer and presented by John Mather in January 1990, just eight weeks after *COBE*'s launch.[1] Boxes represent conservative estimates of measurement uncertainties. The data show no discernable departure from a perfect blackbody spectrum (the curve) with a best-fit temperature of 2.735 ± 0.06 K.

Cosmic background is claimed to support the big bang theory, but Cosmic Background Experiment (COBE) and the Wilkinson Microwave Anisotropic Probe (WMAP) cosmic background survey findings better support the new universe model, as presented in part II of this document. Max Planck predicted spectral distribution of radiant heat energy from any black body in 1901 and is now considered as 'law for energy radiation from black bodies'. The "black body" name applies to all objects and was so named to differentiate between a body's thermal radiation energy versus reflected or self-generated energy.

The amplitude of the Planck curve defines the body temperature. The radiation curve for black bodies at 2.7 degrees Kelvin is similar to other higher or lower temperature black bodies except for amplitude. The cosmic radiation data was found to correlate with that curve very closely. Those results show that the cosmic background is a typical black body radiation source and independent of any origin of the universe concept. The measured dark body radiation energy level is simply a function of the object temperature and distance.

Figure I-3. Black bodies.

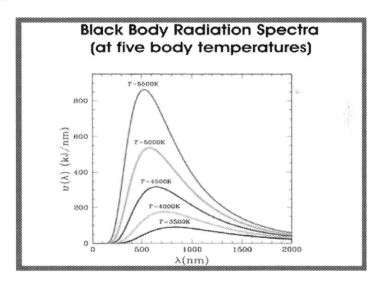

Now, we need to explain the radiation source, one that is consistent with the laws of physics. Microwaves originate from matter and do not just float around in space. A black body is an object that radiated energy from its body temperature and not from reflections. Photons, units of energy, are released from electrons rotating around atomic nuclei. When the revolving electron changes rotational energy level, it either absorbs or releases an increment of energy.

The CMB microwave radiation source necessarily is from dark matter that has been shown to permeate the universe. It has been measured to have a radiation spectrum with the amplitude that calculates to be equivalent to 2.735 degrees Kelvin. The most logical assumption for the radiation sources are/is numerous black body objects that are known to permeate the universe, not just microwaves floating around in space, left over from some imagined process.

Eighty-seven percent[5] of the universe's mass has been shown to be dark matter. It is likely to be a combination of objects (isotopes), some not yet recognized or proven, and helium atoms and simple multiple neutron nuclides. Charts of the nuclides (reference 15) unavoidably automatically have a box position that is empty, a clue to at least part of the universe's dark matter. The empty box fits where multiple neutron nuclides would be, if indeed they do exist, (like helium isotopes but with no protons in their nuclei. These speculated particles would be stable multiple neutron nuclides) (i.e., Mnn).

---

[5] By observing galaxy rotation rates, their observable mass combined with their rotation rate calculates to prove the centrifugal force far exceeds their recognized (from observable) gravitational self attractions. Spiral galaxies would fly apart and thereby self-destruct if they didn't contain a significant amount of dark matter. Dark matter needed to provide adequate mass/gravitational force for stable observed galaxies as observed is eight times more mass than the observed mass. Therefore, dark matter is seven eights of the total, or 87 percent of the galaxy mass. (Paper written by Kenneth C. Freeman, published December 12, 2003, in *Science*.)

And without any protons, so there are no orbiting electrons to produce magnetic fields. Mnn have only mass gravity for mutual attractions; therefore, there is no coalescing via electrostatic and magnetic forces, as do electron-hosting isotopes like hydrogen, helium, and lithium, which coalesce to produce first-generation stars.

However, a portion of the universe permeating dark matter in the CMB necessarily includes some nuclei with orbiting electrons to provide the radiation observed as CMB radiation. Dark matter is speculated to be a homogenous mixture of helium and Mnn.

Because most of the observed mass objects in the universe are in galaxies and galactic clusters, the universal distribution of dark matter is assumed to be similar. Dark matter is known to permeate the universe and possess the mass with the heat that radiates their self (heat) energy in the (Max Planck) microwave spectrum. The strong gravity fields of galaxies and other massive objects would, of course, influence the distribution of dark matter, including multiple neutron nuclide clouds, but not have adequate mutual gravity for capture of Mnn.

This CMB explanation for background radiation energy source is speculative but plausible without any myth. The energy source that provides dark matter energy for its self heat could be—and are likely from reverse radiating annihilation gamma photons from primordial matter annihilations.[6] Forward-radiating gamma wave photons are energy that precipitate into all other mass objects.

---

[6]   Annihilation gamma wave photons are explained in part II.

Photon:

There are numerous photon descriptions, but I have never read a satisfying explanation for their physical description.. By conventional definition; a photon is a unit of electromagnetic energy. Apparently, photons cycle between a linear wave and a corpuscle, and a specific length cannot be concluded.

I think of a photon as an electromagnetic wave that is basically one dimensional[7] and of a finite length, traveling at the speed of light. The energy it contains is transversally oscillating in one or more random directions perpendicular to the one dimension of travel. The finite length of the photon includes one, several, or many wavelengths, as defined by its origin.

Annihilation photons gamma waves wavelengths are only about 0.00002 nm in length, but the annihilation photon lengths are not (yet) specifically definable. The amplitude of the transverse oscillations defines the amount of energy contained in a photon packet. Annihilation photons each contain 511,000 electron volts of energy. Mass equivalent can be calculated using Einstein's e = mc^2 equation. Annihilation photon mass equivalent calculates to approximately 9.109×10^{-28} grams of mass.

Conclusion for CMB Origin;

Dark matter absorption and re-radiation of energy in the spectrum is measured as "black body proven to exist". Dark matter is a more scientifically plausible explanation than to say CMB came from a once unitary nothing and therefore proves something.

---

[7] It is one dimensional, yet observed in many directions from the source simultaneously. Dozens of concepts are illustrated on the Internet.

Problem #6: The Big Bang Theory Limits the Universe's Age to 13.77 Billion Years

That claim is based on the farthest distance and time that anything can be observed. However, the timeline bar chart for our region of the universe illustrates the sum of the proven sequential building blocks, demonstrating the universe is much older than the big bang concept allows (*New Universe Theory with Laws of Physics*).

Accretion times between star burning/fusion serial time increments between time bars vary and are not known for any specific region. However, time increment ranges have been estimated and agreed as reasonable by many astrophysicists. With this bar chart, we prove the part of the universe in which we live is at least about 30 billion years of age, possibly younger and probably older.

It requires at least three ancestor generation/cycles of star formations, burning, supernovae, and recoalescences of supernovae debris to produce the heavy element products in Earth and the sun, (Sol). Our sun and the rest of the solar system are now about 4.5 billion years of age and have about 4 billion years to go before our sun expends its fuel. The shortest possible time to develop the current local system is about 26 billion years if all events occurred expediently as possible. Thirty billion years are shown on the chart. Some cosmologists should be able to develop a statistical estimate for debris coalescence times between first-, second-, and third-generation stars and thereby better estimate the time age of our solar system, which would most likely result in a longer time. The timeline bar chart shows the end-to-end ancestral development stages. The stages are individually described in part II.

Production of metals (heavier than iron) requires a minimum of at least three generations of stars, including the high pressure and temperature fusions within their third-generation stars' supernova explosions. From the universe's initial coalescence of matter to critical mass necessary for the first stars' ignition through fusion and supernovae, which also disperses matter for the second round of coalescence to form descendant critical mass and ignition of follow-on stars to fuse elements into the heavier elements and then produce and disperse via subsequent supernovae.

Matter—again with time—coalesces until third-generation stars achieve sufficient mass for ignition and thermonuclear burning to produce the elements we observe throughout the solar system. Three cycles are required with the heaviest elements being produced within the compression forces of the supernovae as they disperse. The fourth accretion process included production of our star and planets, which also include heavy elements. Reviewing a chart of the nuclides helps one to understand the phenomena processes.

The universe's and solar system's timeline bar chart (Figure I-4) illustrates where we are, assuming the required sequential events occurred in the most expedient possible times. It could require much more or less time for the development of other star/planetary systems.

Note

All of the timeline bar lengths are justified and explained as observed and explained by many astronomers. The four small bars on the lower side of the main time bar illustrate history and future, including the eventual demise of the sun, Earth, and solar system within the Milky Way galaxy. It includes all biological living beings with knowledge and lack thereof. The solar system requires 0.275 billion years to revolve about the Milky Way galactic core. We have revolved sixteen times, and it is estimated we will do so another sixteen times before we become just another white dwarf star.

Figure I-4. Universe's milestone bar chart.

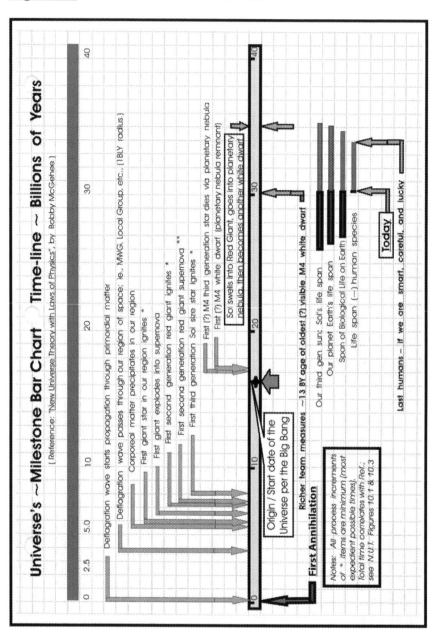

Note the fat arrow on the lower side of the central bar, which indicates when it is claimed the big bang occurred. The "Today flag" indicates the universe's age is greater than or approximately 30 billion years. Also note that time and age is not to be confused with distance dimensions.

Chronological Time Increments on the Bar Chart:
The timeline chart clearly shows that there simply was not enough time for the big bang concept to produce the universe as it is known to be. The fat arrow on the bottom of the bar is when the big bang proponents say the big bang occurred. Even with the impossible "higher than the speed of light" (inflation) phase, as Alan Guth proposed, there was not enough time. The processes before the fat arrow time are essential to produce the universe we see today.

Unanswerable with the big bang is why globular cluster stars are older than stars in the Milky Way host galaxy and yet populate the same region of the universe. These faded white dwarf stars are more than 13 billion years old, which includes no time for three-star ancestral generation cycles of coalescing, burning, supernovae, and dispersion, as are essential for the life and fading by cooling of M4 and M2 white dwarf stars.

Studies of faded white dwarfs in star clusters are continuing. Two of these studies were started early in the decades of 1990 and 2000 by two teams of astronomers, both headed by Harvey Richer of the University of British Columbia. They started by observing the dimmest white dwarfs in the M4 globular cluster. The bar chart milestone identifying this observation is on the lower side of the time bar and is labeled "Richer Team," also labeled "Today" in Figure I-2.

Problem #7: Direct Observations

The preface for Problem #7, *Model of the Universe*, illustrates galactic cluster redshift data observed from the same galactic cluster after a forty-year interval yielded the same redshift. It was hoped that astronomers would recognize there was no acceleration or expansion, but those who read have not yet recognized the negating impact that observation should have on the big bang expansion concept.

The three-astronomer team of Abell, Corwin, and Olowin—who documented their findings in 1959—catalogued the first dataset. Forty years later, the two-astronomer team of Struble and Rood cataloged their measurements in 1999 (*Model of the Universe*).

More recently, since 2010, I became aware of the Spitzer Space Telescope's ability to measure redshift data to six significant figure resolution. The two sets of previous data by Palomar was limited to four-digit figures. I then prepared a Spitzer research proposal for again measuring the redshift of the largest twelve galactic clusters with the thought that comparison of Spitzer six digit data with the fifty-year ago Palomar four-digit resolution could be accomplished with rounding up the older four-digit data, thereby producing two sets of data, one each for the maximums and one for the minimums. I submitted the proposal draft.

To my dismay, I received a reply the next day that stated, "Spitzer does not accept study proposals from non-invited sources." However, about three days later, I received an invitation to submit a proposal. (Apparently, the staff or someone read my proposal draft and recognized it had significant merit.)

Again, dismay and distress, in the process of submitting and coordinating my proposal, I was notified that Spitzer had depleted its cryogen supply and could no longer obtain six-digit resolution redshift data. So I withdrew my proposal.

Shortly thereafter, elation! The Spitzer proposal coordinator called and informed me that the six-digit resolution Spitzer data I was seeking already existed, and various researchers had obtained it. She provided the catalog location/sites. It was cataloged in 2009.

And indeed, it contained six-digit data for the largest galactic clusters as desired, along with six-digit redshift data for several additional galactic clusters. (Presented are two sets of data in Figures I-4 and I-6.)

Comparing the 1959 four-digit redshift data and the 2009 data clearly shows that, on the average, (1) galactic clusters were decelerating and (2) there are only a few clusters with apparent higher velocities (Figures I-4 and I-6). This is direct proof that there is no dark energy and the universe could not have originated in a central eruption. Also, this strongly supports the new universe theory with the laws of physics, as described in part II and also more explicitly in references 1 and 2.

A friend and colleague, Robert E. Farrell, who has a Doctorate of Engineering, degree had previously been involved in several discussions relative to the new concept and agrees with this new universe theory. Dr. Farrell was welcomed to be a contributor, and he helped analyze these sets of data. He used Microsoft Excel and graphed both sets of data. He also converted the redshift data to velocity using Hubble number of seventy-three for this redshift range (Figure I-2.) The fifty-year time-lapse yielded two data plots that clearly display, on the average, galactic clusters are decelerating. Farrell's graphs are presented in Figures I-5 and I-7.

We prove there is no universe-accelerating expansion, but there is continuing growth. Comparisons of 1959, 1999, and 2009 galactic cluster redshift data proves Hubble's 1920s interpretation of Slipher's 1910s measured increase of redshift versus distance was his misinterpretation.

Each galactic cluster occupies and represents its local region of the universe. All observable galactic clusters, in both north and south hemispheres, were cataloged in 1959 by Abell, Corwin, and Olowin, whose data was acquired from PASS, the initial/first-priority study for the then new world's largest 'Palomar' telescope.

The ACO team produced a listing of 4,076 galactic clusters along with cluster vitals, most importantly, redshifts. They cataloged galactic cluster sizes and listed all of those with thirty or more galaxies with redshift and coordinates (right ascension and declination) for each cluster. I downloaded the entire catalog into Excel for analyses.[8] Otherwise, this listing includes all of the galactic clusters in the universe that contain over thirty galaxies.

The nature of galactic clusters is such that clusters are large and the galaxies within each meander about within the cluster family's gravitational field(s). It is recognized that the recorded redshift data

---

[8]   Olowin later cataloged an additional thousand or so southern hemisphere clusters that are not included in these analyses.

from each cluster are from that cluster's individual tag galaxies. Some multiple galaxy cluster-averaged statistics have been excluded from this study because of uncertainties, especially relative to time-lapse measurements. Meandering within individual clusters could cause that cluster's (tag galaxy) redshift data to vary over our forty-year time-lapse period.

Table 1 shows the eleven largest galactic clusters in this survey. These are believed to be the largest clusters in the universe. They continue to grow and decelerate as more rogue stars and galaxies are accreted.

It is important to define some terminology at this point. The term "expansion" assumes space, distance, and velocity among objects are increasing. The term "growing" assumes additional matter and space are being added continuously with time. Tables 1 and 2 show recently acquired data, which demonstrate the universe is growing with no expansion of prior space. These data substantiate the new origin of the universe concept described in part II of this document, originally described in references 1 and 2.

Galactic Cluster Data

Data common to all three catalog listings (ACO, S&R, and Spitzer) are listed in tables 1 and 2 to sort out the older, highest-count clusters for analyses and directional studies. Before Spitzer data became available, all common cluster data from both ACO and S&R are listed in *Model of the Universe*. Also in appendix 2 of *Model of the Universe*, calculations are displayed that show the universe is not accelerating, as the Hubble equation states. There is no discernible forty-year redshift change when comparing ACO and S&R data.

Figures I-4 and I-6 are graphed from analyses of data from the same clusters, which now proves the galactic clusters and universe are decelerating. The deceleration line shown proves no big bang acceleration and is similar to the deceleration lines again illustrated in figure 2 of part II.

Table 1 contains the ten largest galaxy count clusters. This table data has been sorted from all of the 4,076 clusters cataloged to identify and list the largest galaxy count clusters. Their galaxy counts range from 190 to 321.

Table 2 lists the other ACO galactic clusters common to all three special catalogues. Struble and Rood (1999) catalogued data from their survey of some of the same of ACO clusters (listed in appendix 5 of *New Universe Theory with Laws of Physics*). More recently, in 2000 to 2009, very high-resolution (six-digit redshift) Spitzer again measured some ACO clusters listed in tables 1 and 2 (cataloged 1999 to 2011 in the NASA/IPAC Extragalactic Database or NED[9] by the Spitzer team.) The orbiting Spitzer Space Telescope with its cryogenically cooled instruments was capable of high-resolution redshift measurements. To everyone's chagrin, the cryogen was depleted in 2009.

---

[9] Anyone can access the database sources for verification (http://ned.ipac. caltech.edu). Click **Redshifts** on the pop-up page. In the **Object Name** box, insert the ACO Number (e.g., ACO 895). On the pop-up data page, note the redshift is posted with all six digits. This is a wonderful database for exploration and self-education, along with providing proof the objects in the universe are decelerating.

Figure I-4 **(Table 1).** The ten-largest galaxy count clusters with data available for the fifty-year time span including recent Spitzer six-digit redshift.

Data proves universe's galaxy outward flow is slowing						
50 year Z & delta Z						
(1959 through 2009)						
If just one cluster is not accelerating (has no red shift increase) there is no universe expansion.						
Cluster	# Galaxies	RA	Dec.	Z 1959 & 1999	Z 2009 Spitzer	50 yrs Delta Z
ACO ID #	in cluster	(REF 1950)		ACO & SR		1959 to 2009
(Tag galaxy)		(Tag galaxy)	(Tag galaxy)			
10 largest galactic clusters in the ACO catalog with NED data						*...Deceleration:*
586	190	07 29.1	+31 44	.1710	.1701870	-.0008130
2721	192	00 03.6	-35 00	.1140	.1148660	.0008660
1758	198	13 30.5	+50 46	.2800	.2790001	-.0009999
2218	213	16 35.7	+66 19	.1710	.1709000	-.0001000
910	222	09 59.1	+67 24	.2055	.2056220	.0001220
1140	222			.1410	.1410000	.0000000
3558	226	13 25.1	-31 14	.0482	.0477130	-.0004870
2125	230	15 40.5	+66 28	.2465	.2461630	-.0003370
545	234	05 30.0	-11 38	.1540	.1546190	.0006190
665	321	08 26.2	+66 03	.1816	.1810360	-.0005640
				1.7128		-.0016939
	2248		average	.1713	0.1711106	-.0001694
			delta Z average, 10 largest ACOs			-.0001694

These data include not only the largest. These are likely the oldest galactic clusters in the universe. Graph of Table 1 data illustrates the universe's largest clusters are decelerating. Note the zero velocity change line (data sorted from ACO, S&R, and Spitzer 2009 catalogs; graph via Microsoft Excel).

Clusters yield some of their linear momentum as they mix and accrete more matter into the cluster. That is, they linearly decelerate as they acquire more angular momentum. The largest galactic clusters in the universe (on record through 2011) have forty- and fifty-year time-lapse redshift records that prove deceleration [figures I-4 (table 1) and I-3].

Figure I-5. Average deceleration is approximately 2 Km/sec/year.

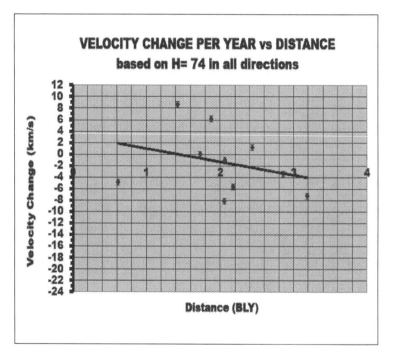

Universe's Largest/Oldest Ten Galactic Clusters

Figure I-6 (Table 2). Galactic cluster data with fifty-year time span and six-digit redshift data.(as of 2012 AD).

Cluster	# Galaxies	RA	Dec.	Z 1959 & 1999	Z 2009 Spitzer	50 yrs Delta Z
Common Clusters in ACO & SR catalogs + NED data (Z > .2) (31 clusters, 3671 galaxies)						
209	158	01 29.5	-13 50	.2060	.2061190	.0001190
223	152	01 35.5	-13 02	.2070	.2062070	-.0007930
528	40	04 56.9	-09 05	.2896	.2901710	.0005710
593	154	07 46.2	+70 04	.2260	.2263170	.0003170
639	135	08 15.1	+68 04	.2910	.2910250	.0000250
732	65	08 55.3	+03 22	.2030	.2027310	-.0002690
895	222	09 53.5	+49 44	.3600	.3604920	.0004920
910	50	09 59.1	+67 24	.2055	.2056220	.0001220
1094	83	10 44.8	+27 47	.2004	.2010150	.0006150
1178	103	11 07.1	+34 52	.2596	.2592370	-.0003630
1224	62	11 18.2	+36 42	.2897	.2903470	.0006470
1299	104	11 29.7	+35 44	.2247	.2240830	-.0006170
1304	80	11 30.1	+35 44	.2131	.2128040	-.0002960
1430	96	11 56.9	+50 04	.2105	.2097820	-.0007180
1525	186	12 19.5	-00 52	.2590	.2586210	-.0003790
1550	167	12 26.8	+47 59	.2540	.2530170	-.0009830
1622	96	12 47.3	+50 06	.2855	.2852730	-.0002270
1785	90	13 42.5	+36 24	.2136	.2139290	.0003290
1878	56	14 10.6	+29 27	.2540	.2543480	.0003480
1929	95	14 29.7	+29 45	.2191	.2188600	-.0002400
1942	138	14 36.1	+03 53	.2240	.2240350	.0000350
1954	107	14 39.9	+28 44	.2480	.2471520	-.0008480
1957	166	14 41.0	+31 25	.2410	.2402740	-.0007260
1961	137	14 42.4	+31 24	.2320	.2314430	-.0005570
2111	148	15 37.7	+34 34	.2290	.2279990	-.0010010
2125	230	15 40.5	+66 28	.2465	.2461630	-.0003370
2246	146	17 00.4	+64 17	.2250	.2249900	-.0000100
2270	49	17 26.3	+55 13	.2377	.2384310	.0007310
2317	186	19 08.5	+68 59	.2110	.2110830	.0000830
2444	76	22 24.8	-24 06	.3240	.3244510	.0004510
2616	94	23 30.7	+05 20	.1832	.1835090	.0003090
	3671				7.4727	-0.0031700
		Average Z =	.2411			

		delta Z average., 31 clusters, 3671 g's	-0.0001023
		delta Z average, 10 largest ACOs, 2248 g's	-0.0001694
		delta Z average., 41 clusters, 5919 g's	-0.0001186

Figure. I-7. Graphical proof is now presented for all the galactic clusters with forty- and fifty-year time-lapse data.

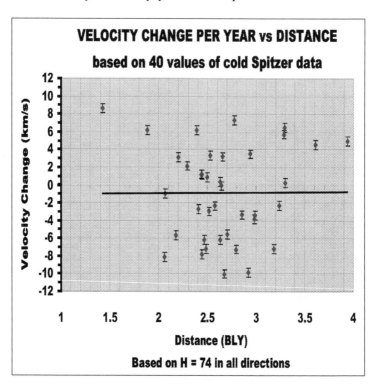

The universe's galaxy clusters are decelerating, not accelerating, as the big bang theorizes (fifty-year velocity change: forty galaxy clusters: Redshift, Z = 0.1 to 0.33)

It's proof that, contrary to the big bang theory, all galaxies are not accelerating. This chart verifies that galaxy clusters are decelerating. (These are all the galactic clusters that are known to exist with six-digit redshift data.) The average deceleration is approximately 1.0 Km/sec/year (graph contributed by Dr. R. E. Farrell).

Further proof will be undeniably conclusive from all galactic clusters with/when more telescopes and spectroscopes yield six-digit data.

Figure I-5 reveals and proves the universe's largest—and therefore the oldest galactic clusters—have decelerated during the time-lapse in the recent fifty-year period. (Ten years should be adequate with six digits at each end of the time-lapse period.) High-resolution spectrometers are needed for time-lapse redshift data over ten-year time increments. We have found and extracted the ten largest galactic clusters in the universe, also with an additional thirty-one clusters from all of the known galactic cluster records available. ACO cataloged 4,076, and it is reported that Olowin added an additional list of about eight hundred from the southern hemisphere. These cataloged clusters include all clusters having thirty or more galaxies. Because we are within 0.3Z of the universe origin, not many more, if any, are expected to be found with over thirty galaxies. The number of galactic clusters with less than thirty galaxies, such as in our Milky Way's 'Local Group' is unknown, and there must be many.

Much work is yet to be done. To map the universe, six-digit resolution Z data is needed for all modern new technology telescopes. **The new currently in process James Webb space telescope direly needs planning for on-going servicing for high resolution (six digit) redshift data acquisition.**

Note

Galactic cluster velocities were calculated using the equation: $v = c$ $[\{(Z + 1)2 -1\} / \{(Z+1)2+1\}] = \_\_\_\_$ Km/sec. Velocity of light was used at 300,000 Km/sec for expedience. Pertinent equations are listed in appendix 6 of *Model of the Universe*. For example, from 1959 to 2009, a cluster at Z = 0.330000 is decelerating at 50 Km/sec/50years.

A fifty-year velocity change graph uses the 1959 four decimal place data, but to extend the resolution, the possible high and low values from which it could have been rounded up, there was 0.00005 added for the high value and 0.00005 subtracted for the low value. For example, reasoning was for a cataloged value of, say, 0.1234. When upgraded via rounding, it could have been as high as 0.12345 or as low as 0.12335. The data was plotted in Microsoft Excel, which has a plot option that produced the velocity line. The velocity reduction relative to our position in the universe for clusters in the range of Z = 0.10 and Z = 0.35 between 1959 and 1999 have been (as stated earlier) calculated from ACO (CDS) cataloged data and Spitzer (NED) cataloged data. For example, from 1959 to 2009, a cluster at Z = 0.330000 decelerated 50 Km/sec per fifty years.

Linear velocity reductions are simply the result of converting momentum from linear into angular, consistent with the laws of physics, which is typical of all fluid flows. The big bang theory provides no explanation for the source of the tremendous amount of angular momentum in the universe. The big bang concept ignores the ever-increasing entropy, consistent with the second law of thermodynamics. Deceleration streamlines were estimated in figure 2 of *Model of the Universe* (figure II-2). A small segment of the deceleration lines are now defined and known.

Problem #8: With the Big Bang Concept, "All Matter Radially Propagates from Unity" and, as Per Big Bang Cosmologists, There Can Be No Asymmetry

Big bang theorists have proclaimed there can be no large-scale asymmetry because, according to that theory, everything came from unity, which they say is now everywhere. Therefore, distribution of matter must be the same everywhere, outside of some small, local sample comparisons. Yet, much asymmetry exists, both large and small scale.

Example One

Figure I-9. Asymmetry (charts and text reproduced from New Universe Theory with Laws of Physics).

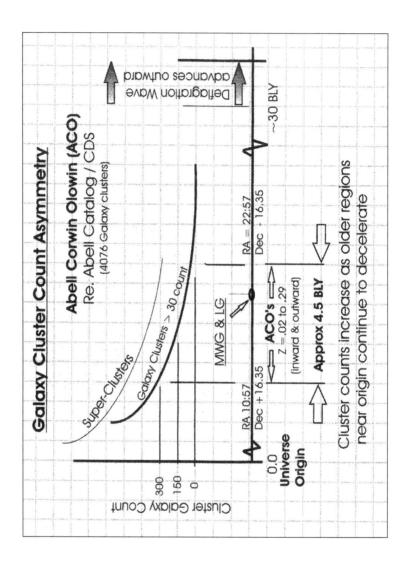

The data compilation by the ACO team have only a few peculiar anomalies, and I did not want to edit (alter or taint) the original listings by ACO contributors, so all ACO data was used as cataloged.

To produce these asymmetry proofs, the entire ACO catalog data from 4,076 galactic clusters was downloaded into my personal computer. The analyses were done using Microsoft Excel.

The WMAP asymmetry survey of the universe illustrates asymmetry for radiation, whether background or foreground. ACO asymmetry data has been available for decades, thanks to Abell, Corwin, and Olowin. In this ACO population graph, the larger clusters (vortexes) are to the right, which indicates the direction toward the older universe. The universe is therefore growing (adding) to the left in this view. There is much asymmetry.

Example 2

In 1957, the ACO team initially headed by George Abell (who died in 1958) used photographic plates from PASS and compiled a list of 2,712 galaxy clusters with redshifts up to Z equal to approximately 0.24, each containing at least thirty galaxies, and included some with over three hundred. In 1989, Harold G. Corwin from the University of Edinburgh (retiring in 2011 from Cal Tech) and Ronald P. Olowin from Oklahoma University extended the list to 4,076. Olowin added dozens more clusters in 1959. This completed the ACO catalog, which now includes the southern hemisphere. (The only region not included is that portion of the sky, which will continue to be obscured from view for the next 62.5 million years by our home galaxy, the Milky Way.)

We do not have a count of the number of galactic clusters with less than thirty galaxies. However, a more recent survey known as the "Galaxy Zoo" may have a partial list for galaxies with quantity numbers for less than thirty galaxies per cluster.

Figure I-10. Cluster population distribution (ACO catalog).

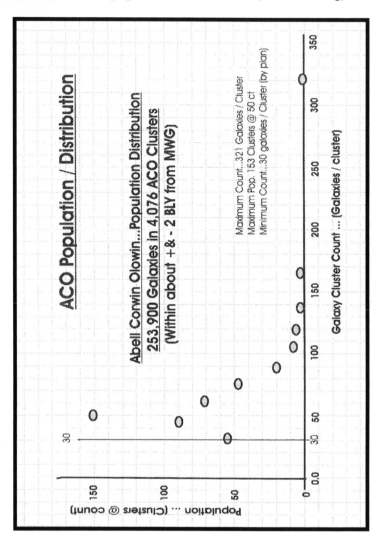

Knowledge gained from figure I-8 states that the largest—and therefore oldest—clusters are fewer in number as they merge and accrete more of the available matter and contain additional galaxies (to the right in this graph).

This asymmetry proves the universe is growing in the opposite direction (from the left), toward, and following the universe's periphery, where more objects are being added by matter conversion from primordial matter. The largest cluster population is as expected of smaller galaxy count clusters and are in the younger part of the universe, to the left in this chart.

An interesting/curious question is why data indicates the total ACO population/cluster count peaks at 153 clusters with fifty count is not yet understood. Our local group (less than twenty galaxies) is to the left of the thirty-galaxy cluster count line. There is much asymmetry.

Example 3
Galactic clusters in the observable sky were cataloged with the world's largest telescope in the 1950s in PASS, the first dedicated project for the world's then new largest telescope, by Abell, Corwin, and Olowin. Non-symmetry of the universe was cataloged and is now analyzed, which further proves asymmetry. The older regions of the universe are logically going to have more time to accrete galaxy cluster members and grow larger. The largest structures in the universe are galactic clusters. The documented clusters were sorted by size from all of the 4,076 ACO clusters. All clusters with more than thirty galaxies were documented with the Palomar Telescope. The largest fifteen galactic clusters were sorted from the 4,076 total and then further sorted by direction. The largest clusters are redshifted in the direction coordinates of 10:57 (Hms) right ascension and +16:35 (Dms) declination (*New Universe Theory with Laws of Physics*). There is much asymmetry.

The following graphs, Figures I-11A and I-11B, are also significant for locating and verifying the universe's origin.

Figures I-11A and I-11B. Fifteen highest count ACO galactic cluster directions:

Right Ascension:.

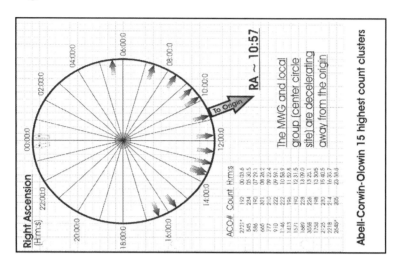

Declination:

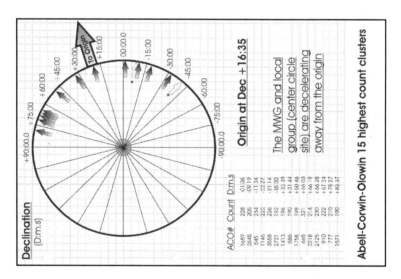

Confirmed asymmetry disproves any unity origin for all matter and space. The arrows point in the direction of the Universe's largest/oldest clusters, likely the origin site.

Problem #9: Voids

The existence of any void contradicts the big bang premise that requires the universe be totally symmetrically populated. Voids are sites where there is/are nothing—no mass, no radiation, nothing. Big bang advocates do not have any explanation for these discoveries. Usually when any new discovery is made, the advocates rationalize and claim it as big bang verification. With most new discoveries, they often say, "This proves the big bang origin," for example, like was often erroneously stated when background radiation was first discovered.

There are at least two voids, and it is not possible to explain even one empty void in a universe from the hypothetical big bang that theoretically erupted and spread matter radially in all directions from a single point to fill the universe's volume. The mere existence of a void is in conflict with the big bang concept of the universe originating and radiating in all directions from a central explosion.

Two major voids have been discovered, and it is agreed that they are very puzzling. A credible explanation is offered in part III that is consistent with physical laws and does not require accepting myth for understanding. It is understandable with only facts and logical analyses.

Some have speculated voids are massive black holes, but voids lack the known characteristics of black holes, as do both the billion-light-year-diameter Eridanus Void and the 500-million-light-year void near Abell Super Clusters.

The two largest voids found to date are:

- A huge void almost a half-billion light-years in diameter and a billion light-years away was discovered in Bootes in 1981 by astronomers Robert Kirshner, Augustus Oemier Jr., Paul Schechter, and Steven Shectman. They reconfirmed the void finding after further surveying the area in 1983. By 1997, sixty galaxies were discovered near that void. Directions to the Bootes void are RA 14 degrees 30min and Dec approximately +45 degrees. Note: Recent reports indicate that, in 2009, radio astronomers report the Bootes Void as 250 million light-years in diameter and about 500 million light-years in distance.

- Now, discovery of the largest void ever was found in 2007 (with info released in 2009) by Lawrence Rudnick, professor at Minnesota University, and his colleagues, Shea Brown and Liliya Williams (reference 12). It is located at RA approximately 3hr 30min and Dec approximately -50 degrees. It is said to be over a billion light-years in diameter and about 10 billion light-years away from our Milky Way galaxy. The Rudnick Void is also referred to as Eridanus Super Void. It is located in the Eridanus constellation, which is extremely large, located between 1 to 5 hr RA and between 0 to -60 degrees declination. In early June 2014 correspondence, Lawrence Rudnick said the void has not yet been confirmed, and he doubts it exists, but there is something quite odd about that position in the sky. More recent correspondence indicates the National Astronomical Radio Observatory (NARO) may have verified the void.

- Could that Rudnick Eridanus Void be the universe's origin site? Evidence indicates yes. This will be discussed in part III.

Problem #10: Spherical Growth, Gravity Waves, and CMB

This isn't just proof. It is simply logic and realism. The equation for the volume of a sphere is $V = 4.189\ r^3$. The relationship between a sphere's radius and its volume shows that, every time the radius (r) doubles, the volume (V) increases eight times. The big bang believers say the universe is now 13.77 billion light-years in radius. The equivalent diameter is 27.54 billion light-years. Some big bang believerssay everywhere is the center; therefore, it is implied and has been stated by some astronomers that the diameter is the same as its radius … not. And impossible!

The big bang concept says the universe is growing in all directions at the speed of light (both diameter and radius.) Its growth is said to be increasing with the energy to accelerate all the mass in the universe, powered by so-called dark energy. Dark energy explains hypothetical and other illusioned or scientifically unexplainable, imagined phenomena. This is not to be confused with dark matter, which has been proven to be over three-fourths of the universe's mass. None of those dark energy ideas can be supported by any of the laws of physics or other proven facts.

How much proof do we need to show all the matter and energy in many billions of galaxies, many trillions of stars, and many trillions of cubic light-years of space could not have come from a single, dimensionless point called unity? Where did all of the angular momentum in the universe originate?

Part I Summary and Conclusion Laws of physics violations have always made scientific thinkers uncomfortable with the big bang. Part II is a credible explanation for how the universe began that is consistent with proven laws of physics. Rationalization that a discovery proves a previous conceived theory is usually only that, dogmatism; (CYA rationalization, and nothing more). CMB, as first observed in 1947, proves there are background radiations and nothing more. Gravitational waves, observed in 2011, proves there may be gravity waves and nothing more.

Those discoveries alone are significant. The cloak-and-dagger teams that worked in close secrecy for two or three years with their data so competitors would not get credit was cynical motivation. The teams were more interested in their egos and notoriety than contributing to mankind's knowledge. The joke is on them because it is now proven that there never was a big bang. All the rationalizing about how the gravity wave observations explain the big bang is simply not true. It now appears that no gravity wave was observed. Just like all the fabrications about inflation, it's more CYA to evade the embarrassment of facing up to all the stories told to defend and explain how the big bang worked.

But the big bang never happened. There is/was no inflation. You too can prove with certified data that the ten largest galactic clusters in the universe are decelerating[10] at 2.0 km/sec/year while

---

[10] Anyone can access the database sources for verification of proofs (http://ned.ipac.caltech.edu). Click **Redshifts** on the pop-up page. In the object name box, insert the ACO Number (e.g., ACO 895). On the pop-up data page, note the redshift is posted with all six digits. Also, the Abell and ACO galactic clusters were downloaded to Microsoft Excel and sorted in several different ways: galaxy cluster size, size and direction, size and redshift.

thirty typically large galactic clusters are decelerating at 1.0 km/sec/year. Data analyses are based on a fifty-year time period between observations (1959 and 2009). Galactic clusters used are between z = 0.2 and z = 0.36. The results are not just someone's interpretations.

Part II: Alternative Universe Origin

Introduction

Part II presents this now proven alternative origin that was first published in 2005 as "New Universe Theory with the Laws of Physics." In this text are analyses and assessments of new data that has been acquired and available only since 1999. Now, it has been verified with additional data that has only been available since 2009. This 2014 confirmation of concept update is now presented as part II of *Big Bang Problems: How the Universe Began.*

The laws of physics are not just words and phrases put together and voted on by some elite group. They are physical constraints proven many times and never disproved. It is also to be recognized, just because mathematical equations can describe a phenomena, real or imagined, that equation does not prove validity or reality. For example, revisions have been recommended for the Wikipedia encyclopedia entry that presents mathematical explanations of the big bang, which we prove many times over cannot be valid.

Through the ages, philosophers, poets, theologians, and astronomers have speculated on the origin and destination of the universe. As stated by an unknown philosopher, "The universe in which we live, as far as we know is unique and we have no opportunity to check any hypothesis by comparison with other examples."

For the past several decades, some have accepted the big bang as the universe's origin. We all must recognize that credible and respected astronomers in their day (1910s through 1930s) developed the big bang concept. Even now, into the 2010s, many astronomers have continued to accept it. Cutting the old developers some slack, they could only develop a concept with their limited knowledge. High-energy sources—such as matter and antimatter annihilations— were not known until the 1930s.

It also seems to be a human characteristic to become bigoted and polarized in favor of prior beliefs. Peer review requirement by publications, libraries, and some organizations successfully prevent bogus ideas from being published. However, it also perpetuates dogma and prevents advancement of understanding. By definition, belief is to accept without proof. After so many years, it is no wonder the big bang theory is defended such that all other and new ideas have been considered bogus, even though new ideas are more consistent with laws of physics and proven facts. The big bang concept is not consistent and requires acceptance of hypothetical, improvable, and disproved phenomena.

At a local astronomy club meeting, I pointed to several reasons why the big bang cannot be valid. A retired preacher turned to me and asked, "What difference does it make?" My answer was, "The awe of the nighttime sky inspires scientific-thinking people to want scientifically creditable origin explanations."

Before continuing, please refer again to the pertinent laws of physics and nomenclature listed at the beginning of this book.

Verifiable Universe Origin
"It is more logical to assume that everything developed from space and matter that already existed than to assume everything came out of a single point" (reference 3). It is generally agreed that, by far, most

of the observed matter in the universe are hydrogen atoms. Several cosmologists have estimated that, if the total matter (mass and energy), not including dark matter, in the universe were distributed throughout all space as hydrogen atoms, its distribution would be somewhere between one and eight atoms per cubic meter.

Dark matter is not included in their estimates. Dark matter is measured by its gravity, and studies now indicate it to be about eight times more plentiful than what is conventionally observed (reference 4). Dark energy is mythical and hypothetical. Data presented in part I provide proof that dark energy does not exist.

Because estimates of total matter density vary over a wide range of several factors, the primordial particle populations are referred to as "about," and precise reconciliation of primordial versus corporeal universe mass is not necessary or attempted in this text.

Much of part II repeats previous theory presentations, but now, much of it has been verified, and new discoveries support it.

Interpretatiown of Redshift versus Distance Data

Figure II-1 Red Shift vs Distance: Spud Truck Analogy

Several analogies are presented in *New Universe Theory with Laws of Physics*. (Ref1)

This analogy shows a sketch of a truck spilling potatoes as it progresses up the road.

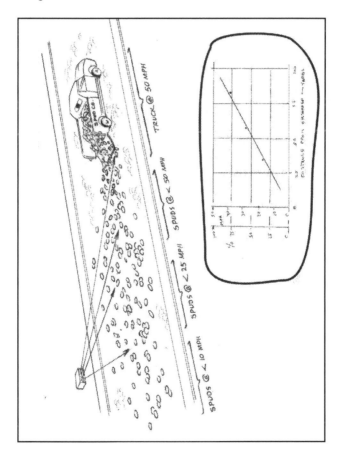

Speeds of individual spuds are measured from the roadside with a Doppler device (radar). The potatoes actually start decelerating the instant they fall from the truck. They continue rolling forward, and their linear momentum is converted to angular momentum as they bounce and roll down the road. The graph insert shows the spud speeds (miles per hour) are faster as the distance (yards) increases from the radar speed monitor at roadside.

A person looking only at the data could easily misinterpret that the potatoes are accelerating as they move away from the observer. The Hubble line type graph appears to show galaxies accelerating with distance. This is how the redshift data for galactic clusters has been misinterpreted as accelerating. This is how the big bang idea evolved. The following graph presents a model that fits the redshift data and explains how the universe is growing while corporal matter of the universe is decelerating with time.

Developing the New Concept
James C. Baker, a Boeing rocket engineer, said, "It is more logical to assume all space was full of something that converted to the matter in the universe than to assume everything came from a single point."

First, it is recognized that the laws of physics were always valid and always will be, as they have been proven many times in scientific experiments and never disproven. The first law of thermodynamics concludes that matter cannot be created or destroyed. Therefore, primordial matter and universe matter are necessarily equal in mass and density.

Next, we needed to identify an extremely high-energy source that could provide the widely distributed conversions of large quantities of mass to energy and back to mass that could gravitationally coalesce and clump. The only known such source of mass to energy is via matter and antimatter annihilations. This appears to be the only answer and fits the need for explaining the universe origin.

Next, a viable concept for widespread primordial matter with a stable geometrical arrangement needed to be deducted, which could be uniformly spread throughout primordial space and also be stable. It is well known that positroniums in the laboratory are very unstable and short-lived. It also is well known that the universe is permeated with extraneous radiation waves: gravity, electromagnetic, electrostatic, magnetic, and gravity. However, the primordial universe is free of all of these. If primordial positroniums were in uniform hexahedrons, they would be equally spaced. Additionally, positronium components (electrons and positrons) would need to be rotating and orbiting in phase and in synchronization with all other primordial positroniums. During an annihilation, typically two gamma rays (photons) are released. From each annihilating positronium, two 511,000-electron volt photons radiate in opposite directions.

With this understanding and reasoning, we proceed.

Now consider primordial space as occupied by the simplest atom in the form of stabilized positroniums (electron and positron pairs). In this model, all primordial space is theorized to be fully occupied with steady state positroniums at each vertex of equilateral hexahedrons. Spacings are stabilized by synchronized rotations, including electrostatic attractions/repulsions. These would be spaced at a separation of about five centimeters in all directions to infinity. This high degree of order would represent the lowest level of entropy for the primordial universe. In primordial space, positroniums could be and are stable where there are no extraneous gravity, electromagnetic, or other waves to disrupt spacing and synchronization. Each para-positronium has the mass of one-nine hundredth the mass of a neutron.

It should be noted that positroniums in the laboratory experience self-annihilation in far less than one second. Extraneous energy and force waves permeate the universe. Also, ortho-positronium is the name for various (other than a single pair) combinations of positrons and electrons. Those are products of thermonuclear and fission processes, as occurring in supernovae and man-made reactors. Positronium types other than para-positroniums are not primordial matter.

Note: A multibillion-dollar research facility for study of positrons and electrons exists in Beijing, China. It has received upgrade funding from seven nations, and it is known as the Beijing Electron Positron Collider laboratory, or BEPC. The upgraded facility returned to operation just this decade.

Logically, the universe started somewhere in the limitless sea of primordial matter, a matrix made of an equilateral hexahedron positioned para-positroniums. Mutual gravity, electrostatic, and electromagnetic attraction and repulsion stabilize the positroniums. For an unknown reason, one of the positroniums experienced self-annihilation, and the universe began.

Although this theory does not exclude the possibility of multiple initiation sites, an initial self-annihilation at a single location would be sufficient to initiate the chain reaction of annihilations that propagated in all directions into an unlimited matrix of positroniums.

Origin of Universe's Matter

Annihilation photons are one-dimensional, finite length, sinusoidal electromagnetic submillimeter wavelength energy packets radiating outward. Forward-radiating photons are a few centimeters in length, and when they overlap, they combine to increase in intensity. When the combined photons reach sufficient intensity, they precipitate into subatomic particles, which ultimately combine and convert into all the matter (mass and energy) that we detect in the observable universe. This multistage process has been coined as deflagration, and it is occurring in a wave that is propagating outward at the speed of light from the site of the initial annihilation.

Gamma ray photons that radiate in directions other than forward do not overlap sufficiently to precipitate into mass and therefore remain as energy packets that ultimately provide energy (heat) to mass objects. Total mass of dark matter has been measured to be approximately 87 percent of the mass of the universe. Dark matter particles absorb the energy packets and radiate heat energy to become what was discovered in 1947 and labeled as CMB background radiation.

Microwave radiation has much longer wavelengths than gamma ray annihilation photons. Dark matter in this concept is now theorized to mostly be multiple neutrons, hydrogen, and helium nuclides that, except for heat radiation and gravitational effects, are otherwise undetectable. Forward-propagating photons combine and produce mass in a manner similar to the processes as described by Steven Weinberg in *First Three Minutes* (reference 5).

Precipitated mass objects, including stars, galaxies, and galactic clusters, follow the cascading annihilation wave outward but rapidly decelerate from the ongoing wave via transfer of linear momentums to angular rotational momentums. Their outward direction is separating from and following behind the annihilation wave that continues at its near light speed velocity. The objects rapidly decelerate and then continue deceleration forever via transfer of linear momentum into angular momentum and mixing.

Development and Growth of the Universe
Following the matter transformation from energy to mass using the reasoning as presented in Weinberg's book and with near light speed wave propagation, the deflagration wave requires a thickness of light-years, not just three light minutes, as Weinberg initially conceived.

He accepted—or did not object to—the big bang explanation and inflation as suggested by Alan Guth as needed to provide/explain what would be required for process time (reference 6). The impossible faster-than-light speed Guth proposed as inflation has been tentatively accepted by many, but it is considered as bogus. NASA accepted and published their version of the point source, including inflation in their chart, as presented in Figure I-1, (page 12).

After the past 30 billion years, the deflagration wave appears to be only 15 billion light-years beyond the region where we reside. This false appearance is simply because that is where the new objects were when the light from those new objects started radiating light, which requires another 15 billion light-years to travel back toward us. The stars were forming in and following behind the deflagration wave. However, it is now about 30 billion light-years beyond us, and it will require another 30 billion years for light from the newest objects to reach/return to us. Thus, the universe diameter is about 60 billion light-years (figure II-7).

Due to the limit of the speed of light, distance to the deflagration wave and newest objects will always be twice what we can see with light waves. The universe's growth is continuingly occurring in the universe's evolution, growth, and continuing development. Figure II-2 illustrates what has and is continuing to progress.

Several Messier object photographs illustrate some of the universe's evolutionary processes. Planetary nebula M14 is the inflating result of a star (similar to our sun), which has burned for 8 billion years and no longer had enough fuel/mass gravity to continue confining the fusion process. When it inflated itself apart, it shed some of its mass into a ring that is called a planetary nebula. In the center of that Messier object is a white dwarf star like those in globular cluster M4, the faded white dwarfs that allowed Richer and team to measure 14 million years cooling time, for those that are still faintly visible, a significant time increment in the universe's age.

Gravitational congregating of stars precipitated out of supernovae debris from earlier-generation giant stars, along with other matter, to coalesce into galaxies, some similar to our Milky Way galaxy. Messier object photographs available on the Internet are a set of astronomical objects first listed by French astronomer Charles Messier in 1771.

Deflagration Wave

The first primordial matter positronium annihilation triggered the cascading of annihilations in the matrix of hexahedron-spaced positroniums. The progression of positronium instability results in positron and electron contacts, and their instantaneous annihilations release 900 electron volts of energy in each of two photons. Propagation of positronium orbital collapses travels near the speed of light, and photons radiate in all directions. Those that radiate forward build in intensity, and as the intensity becomes adequate, mass particles precipitate. These particles' velocity-enhanced gravity attractions induce coalescences. As larger particles are fused via collisions, they collect and accrete until mass of objects become adequate to become stars. Each particle collision and near collision results in rotations and mutual orbiting, thereby decelerating the objects by transfer of linear momentum into angular and rotational momentums. Objects never collide precisely center of gravity to center of gravity.

Figure II-2 is a cross section view of a segment of the spherical transformation (deflagration wave) that is progressing outward into primordial matter. It is thought to be at least a few light-years thick, consistent with the first and second laws of thermodynamics. Matter (mass and energy) cannot be created or destroyed, but it can be converted from and to other forms. There are many forms of energy and many forms of mass. The wave contains the subatomic processes described by Steven Weinberg (reference 5) without impossible faster-than-light velocity, as Alan Guth described (reference 6).

Figure II-2. Deflagration wave processing of primordial matter into the universe's corporal matter.

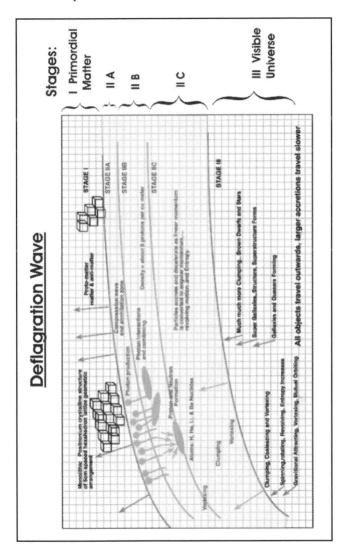

The deflagration wave identifies matter in three stages: primordial matter and transformation processes within deflagration wave. And within the deflagration wave are three sub stages: annihilations cascading, photons transformations into mass particles, and mass particles coalescing into adequate quantities to start fusions and then become stars.

Dividing the deflagration wave into these three sub stages is arbitrary, but these regions are expected to physically exist. Stages are not expected to be clearly differentiated. Linear momentum deceleration would occur throughout, and rapid deceleration occurs beyond where velocity-enhanced gravity decreases with velocity reductions. The first stars come into existence at high velocity, as the Adam Reiss team measured in 2006. Deceleration rates have been calculated from 1959 and 2009 cataloged galactic cluster redshift data.

The first objects start at near light speed, but their linear speeds are only about 0.3 c after only 100 million light-years. Deceleration continues as mass objects collect into groups and grow into clusters of stars, and galaxies form. Deceleration rates continue as galaxies collect into clusters, but deceleration eventually must stop as the velocity becomes closer to nil, but never to zero.

Universe's Growth

Figure II-3 illustrates how the universe grows. It illustrates the vertical axis with three rulers: velocity (km/sec), redshift (Z), and percent of light speed (% of c), all versus the horizontal axis of distance, which is calibrated in billions of light-years. The deflagration wave is depicted across the top of the chart with a series of squiggly, short vertical lines. It continues relentlessly at light speed. Deceleration stream lines of precipitated objects are indicated and based on estimates made before recent calculations. Now we know the slope of these rapid deceleration lines are steeper than as drawn in this figure (II-3).

We now calculate these deceleration lines from approximately 0.99 c down to approximately 0.2 c to extend over a time of only approximately 200 million years. The deceleration line curves to the right are yet to be measured but are presented as estimated in figure II-3. (The vertical grid lines are spaced at 1.25 billion light-years apart.) The deceleration stream lines are now calculated to be within 0.1 billion light-years of vertical from the deflagration wave down to approximately 2 percent of c. That is almost vertical on this chart. However, 0.1 to 0.2 billion light-years is a long distance. For perspective, 1 billion light-years equivalent is approximately 10,000 diameters of the Milky Way galaxy.

Figure II-3. Universe growth (from *'Model of the Universe',* *Reference 2*).

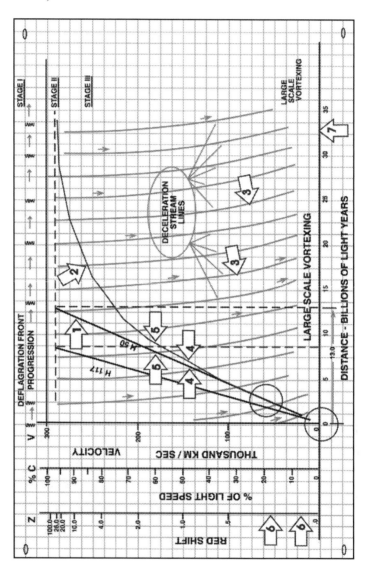

The Hubble line currently used by astronomers is between the two shown on figure II-3. That Hubble line number is seventy-eight. Most astronomers assume the Hubble number line as a straight line of acceleration for all universe matter. It is assumed to extend as a straight line to redshift of infinity. Figure II-2 shows two Hubble lines, which are actually pseudo-gradient lines across deceleration lines. Two lines are shown (117 and 50) because they represent the opposing direction gradients. They are two extremes because of our off-center location in the universe by about 10 billion light-years (part III).

The realistic gradient line shape is curved (not straight) and extends to and becomes tangent to the deflagration wave-universe interface. As time progresses into the future, the tangent point of the gradient line and deflagration wave will also progress farther away from the current contact location. This corrected Hubble line is, of course, curved because we are now known to be several billion light-years from the origin and is based on observations of galaxies in all directions. Realistically, this curved line is a gradient line, not a radiation line as believed by Hubble and other big bang believers.

Deceleration lines near their lower end (approximately 0.2 z) become indefinable and asymptotic to the base line, as would be expected as all linear momentum could not and never will be totally and completely absorbed into angular momentum. Even though there is no accelerating expansion, there will always be slight outward growth. ("Outward drift" is a more definitive term.)

The first two analyses of forty and fifty years of time span velocity change of specific ACO galaxy clusters were done in 2011, providing the first definition of the universe's deceleration rates.

Transfer rate of linear momentum to angular momentum will differ at redshift values other than these between Zs of 0.1 and 0.33 for which data is currently available. See figure II-2 between the two #6 fat arrows (*New Universe Theory with Laws of Physics*).

The analyses to date, presented in figures I-5 and I-7, have defined the initial deceleration calculations. Complete deceleration lines slopes can only be defined in the future with much more/additional six-digit redshift resolution data for galactic clusters at higher and lower Z values.

Allowing present-day astronomers and astrophysicists to return to scientific thinking is a paramount reason why we urgently need six-digit resolution capability included on all spectrometers for use with new technology telescopes. Otherwise, additional time-lapse between measurements will be required, and that will introduce uncertainty as to the clusters' precise locations and representation because their tag galaxies wander and meander within their cluster. Their location within their parent cluster will change with time (uncertainty).

The universe's age is defined by summing of the chronologically sequential increments of time as required to produce the component parts of the universe. The bars are individually substantiated in many references. In total, it is referred to as the "timeline" because of its realistic communication value. This graph is repeated from part I and references 1 and 2. The graph is based on known matter evolution in the universe.

Figure II-4. Universe's age ... Timeline.

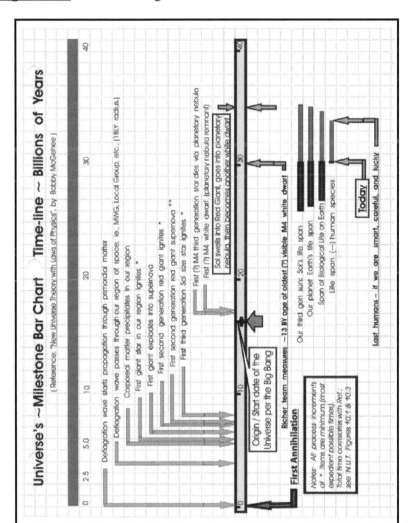

This chart is also included in proof of how the universe's development is inconsistent with the big bang. Yet it is totally consistent with the deflagration theory and the sum total of observations by numerous astronomers and cosmologists. Thanks to all!

Harvey Richer (reference 7) of Canada's University of British Columbia and M4 globular cluster study team measured the age of faded white dwarf stars to be 13 billion years. We know most white dwarf stars are the remnants of third- and fourth-generation stars similar to our sun that have had an 8 billion-year burning phase before starting 13 billion years of cooling. Those stars are descendents of multiple-generation stars, each with ancestry times of their own.

The timeline shown includes the nominal time for our ancestor stars forming and burning before they started cooling. We and our star "Sol" are about 28 (23 + approximately 5) billion years of age. The accretions, nuclear fusion phases, and supernova cycles—plus dimming time through radiation cooling—prove the universe is at least 28 billion years old. That is only true if each of the times between process phases occurred as expediently as possible.

When viewing the chart, it must be remembered that the timeline presented applies only for our location. Every location in the universe has its own timeline from its origin to its present and future, even though the universe's development and growth time is about 28 billion years. We can only see about half the distance to peripheral stars because the light from stars beyond about 15 billion years has not yet reached us. There is an insert arrow to identify where everything began 'according to the big bang idea'.

Bar Chart Increments

Note that time and age are not to be confused with distance dimensions.

Chronological Time Increments on the Bar Chart

This bar chart demonstrates the big bang theory does not allow sufficient time for the universe development events that are known to have occurred. Accepting the new theory—and after the first annihilation occurred—gravitational and centrifugal force imbalances triggered adjacent positroniums annihilations that cascaded in chain reactions. Then the deflagration wave (part II) advanced through primordial positroniums. Precipitations of mass particles ensued with outward velocity to be rapidly consumed into angular momentum. Estimated time increments are listed at the end of each of the following timeline increments.

The universe's matter production deflagration wave reached our region and continued beyond (approximately 3.5 billion years). Based on thinking incorporating the recent Rudnick Void discovery, this may be larger, 3.5 to 10.5 billion years.

Within the wave thickness, primordial positronium matter is continuously converting into mass particles and objects as the wave travels onward and outward. Mass particles and objects precipitated into our region of the universe as it passed through (1.5 billion years), continuing through velocity-enhanced coalescence and clumping of precipitated mass objects, produced stars, including the most senior ancestral giant progenitor stars of M4 white dwarf stars (0.3 billion years).

Fusion burning of fuel within these most senior giant stars was rapid and continued until they exploded in supernovae, which disbursed material throughout the region (approximately 0.8

billion years). Time is required for the supernovae debris to collect gravitationally into star masses (approximately 0.8 billion years).

Debris collected into second-generation, rapid-burning stars and fused some of the lightweight nuclides into nuclides limited to the weight of iron. Second-generation star supernovae again dispersed that material by supernovae explosions. Heavier than iron element nuclides were formed by and during the extreme pressures and temperatures within and during the brief supernova explosions (0.25 billion years).

Some of the second-generation giant ancestral progenitor supernovae debris collected and accreted into third-generation stars, some with masses that are similar to our sun, between 0.9 and 1.44 solar masses (approximately 0.6 billion years).

Third-generation stars acquire sufficient mass and self-gravity to ignite, and their fusion processes continue for 8 to 10 billion years, like in our sun. Near the end of the fusion life of these third-generation stars, they swell in size as the fuel reserves are significantly reduced, resulting in less gravity to maintain compactness. They swell, cool, and then become red giants (approximately 8 to 10 billion years).

At the end of their red giant phase, fuel reserves are depleted, and these stars pulsate between nuclear fusion cycles of "off-on-off." Mass debris is expelled at each cycle of above and below critical mass pressure/density stage. Material is thrown off at each cycle, producing planetary nebula (rings of glowing debris) while ultimately leaving a dormant white-hot white dwarf star at the core (0.01 billion years).

White dwarf stars are totally dormant, no longer having mass sufficient for their self-gravity pressure to support fusion. They cool slowly only by thermal radiation. Cooling to non-incandescence requires about 14 billion years (reference 24).

Total universe evolution time to our current age equals 26 to 32 billion years. From Earth, revolving around our star (Sol), we can observe faded white dwarfs in globular clusters revolving around in our galaxy, Milky Way, that have gone through all of the time increments listed following the timeline chart (repeat from figure I-4). Our Sol will ultimately do the same. Sol and the solar system are estimated to now be between 4 and 5 billion years old, hopefully with 4 or 5 billion years before Sol starts its red giant phase prior to producing a planetary nebula plus becoming another white dwarf.

Most cosmologists recognize our solar system to be younger than the M4 globular cluster stars, which contain and include many third-generation faded white dwarfs. Yet the Milky Way galaxy constituent stars evolved into existence in the same manner, as did many, if not all, white dwarf stars.

How the ages of these star systems are older than the Milky Way host galaxy's stars and are different is logically explained in the new universe theory (*New Universe Theory with Laws of Physics*). As stated above, our younger third- or fourth-generation sun has only lived about half of its 8 to 10 billion years of fusion life. Note the four bars below the overall universe bar illustrates the total lifetime for our sun, Earth, life on this planet, and human existence.

The lowest bar shows the maximum potential life span for our human species. The fat arrow below and to the left of the center of the universe's time bar indicates when the big bang proponents said the origin/start date of the universe began.

On a spatial distance scale, we always will be slowly decelerating outward from the universe origin. Our physical location (radial direction and distance) in the universe is currently estimated (see Voids, 2007 discovery) to be between 3.5 and 17.5 billion light-years from the origin site.

Our galaxy will continue to circulate in and among the local group in the three-dimensional sea of space. Our outward velocity from the universe's center will continue to decrease ad infinitum as vorticity and entropy increase to consume energy from the outward linear momentum of our local group. In the process, our local group will possibly merge with other galactic clusters.

Our species lifetime, if we are lucky, we may have possibly 3-plus billion years before our species, time, progress, and knowledge expires without a trace. It is simply reality. Following Sol's planetary nebula phase, the only remnant will be the sun as another fading white dwarf. It will become a totally dark carbon/ diamond crystal after an additional 13-plus billion years of cooling. Our descendents should not be denied the opportunity to advance our civilization with additional knowledge because of religious prejudice and dogma.. Wars must cease. It is time for humans to quit bickering like little children, each claiming his or her god, religion, and way of life is the only valid one. All is at risk! Let us not destroy what may be the Universe's only intelligence!

On the time scale, we are about 28 to 32 billion years from the universe's origin and start point, but our spatial location is much less. Recent observations (2000–2008) indicate there may be about twenty-plus galaxies revolving and circulating in our local group. Never the less, our local group of galaxies does not yet qualify to be on the ACO list.

Timeline Chart Conclusions

The timeline can be thought of as the fourth dimension that goes along with the three dimensions of space. A unique timeline of events extends from every object in the universe back to the site where the first annihilation occurred. Our milestone bar chart reveals the thread through time from the beginning of the universe, the first positronium annihilation conversion of primordial matter to universe matter. This chart also shows time through and beyond the demise of the last living/biological things on Earth and in the solar system.

The timeline chart clearly shows that there simply was not enough time for the big bang concept to produce the universe as it is known to be. The fat arrow on the bottom of the bar illustrates when the big bang proponents say the big bang occurred. Even with the impossible "higher than the speed of light" phase, as Alan Guth proposed, there was not enough time. The processes before the fat arrow time are essential to produce what we see today.

Unanswerable with the big bang is why globular clusters are older than stars in the Milky Way host galaxy and yet populate the same region of the universe. With the new theory, this is no longer a mystery. It is plausible, logical, and consistent with this origin of the universe concept. These faded white dwarf stars are more than 13 billion years old with no time allowed for three-star generation cycles of coalescing, burning, supernovae, and dispersion, as are essential for biological life.

Several studies of faded white dwarfs in star clusters are continuing. Two of these studies were started early in the decades of 1990 and 2000 by two teams of astronomers, both headed by Harvey Richer of the University of British Columbia. The milestone identifying that observation is on the lower side of the time bar and labeled "Richer Team" and labeled "Today" in figure I-4.

Deflagration Process
Laws of physics are included. Figure II-5 illustrates how mass, force, and velocity relate near the speed of light. For reference, fundamental Newton and Einstein physical relationship equations are inserted. This figure is intended to show what occurs near the speed of light where gravity and relativity maximize..

Ponder that Figure II-5which combines the proven laws of gravity (Newton) and relativity (Einstein) into one nomographic to illustrate the process from where outward-radiating photon intensities become adequate to start ($e = mc^2$) precipitation processes that produce corporeal matter (energy and mass) in the universe. Deceleration occurs as linear momentums are consumed by collisions and near collisions into angular momentums of coalescing and mutual orbiting mass particles.

(Indicated for reference only are the velocities where the gravity between any two like objects would correspond to quark star and neutron star surface gravities.) Astronomers have catalogued several neutron stars, and at least three objects identified are believed to be candidate quark stars. Any objects that increase mass via velocity or accretion beyond the quark star surface gravity disappear as black holes.

Studying and pondering the deflagration process graph leads to the conclusion that there could not be any faster-than-light speed inflation, as would be necessary for a big bang. Also, the mass of the universe would have collapsed into a black hole before it got out of its Schwarzschild radius starting gate. Indeed, accepting the big bang concept requires accepting myth and hypothetical phenomena that are far outside the laws of physics.

Bobby McGehee

Figure II-5. Deflagration process.

Notes in the graph explain the scope of universe processes.

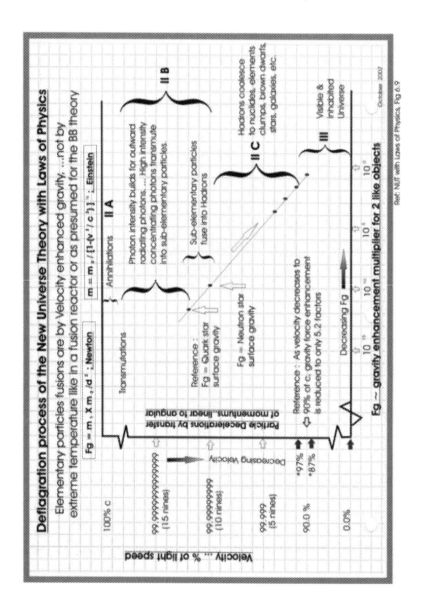

Figure II-6. Ongoing spherical growth.

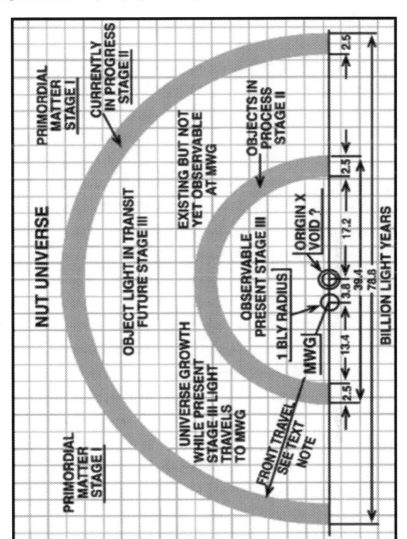

A major change from the initial new universe theory (*New Universe Theory with Laws of Physics*) is the distance between "Origin Void" and our Milky Way Galaxy (MWG). It was not known until the year 2013, but we now know it is farther than 3.8 billion light-years. The observable stage II bands are necessarily centered on the MWG.

Figure II-7. Model of the universe.

This figure serves as the book cover for *Model of the Universe*. It provides a cross section view of the universe. As explained in *New Universe Theory with Laws of Physics*, the universe is developing and exists in three stages:

- Stage I is primordial matter made up of hexahedron lattice/matrix of positroniums.
- Stage II is the transition zone labeled as the deflagration wave, where annihilation photons transform through merging, accretions, and transformations until fusions produce particles that accrete into objects (stars) that produce/emit radiation.
- Stage III is the observable universe (currently observable distance limited to the distance where light from light-emitting objects can reach us, accepted at about 13.77 billion light-years).

Stage III is where matter is and will always continue to evolve following repeated generations of accretions, fusions, and supernovae. The deflagration wave (stage II) travels outward at near the speed of light in all directions (spherically). Light-radiating objects beyond the smaller sphere are not yet visible to observers inside and closer to the initial annihilation and deflagration initiation point. Logic suggests Origin X in Figure II-6 is the site of a void.

To be scientific and valid, all processes and concepts must be consistent with the laws of physics, be provable, and include no myth. Before the universe existed, all space included space and primordial matter. Primordial matter is thought to be positroniums stabilized by geometrically and uniformly spaced, rotating at hexahedron vertices. This primordial matter and geometry is the only scientific arrangement and geometry known that could provide the mass, energy, and space necessary for the universe's development totally consistent with the laws of physics and produce the universe as observed without myth.

The only part of figure II-7 that we cannot visually verify is the matter in the space between the chart rings. The two rings depict the deflagration wave at two different times. It is at the outer ring location at this time, and the inner band is the limit within which we can see matter. Light from objects beyond the inner ring has not yet reached us. We have no reason to suggest that the deflagration front is not continuing to progress outward (distance), as we have proved it has and is producing objects beyond the past 13.77 billion years (time).

The volume inside the inner circle illustrates the universe as we see it. Outside the larger circle, the pre-universe primordial matter is simulated by the hexahedron cubes that are presumed to be everywhere outside the universe. Primordial matter is theorized to be a matrix of para-positroniums suspended and stabilized in primordial space by the dynamics of their rotations, mutual gravitational attractions combined with their centrifugal forces, and electrostatic attractions and repulsions. Synchronization provides a structure reservoir for the matter to antimatter annihilations energy source, unsurpassed by any other known possibility.

A single spontaneous annihilation at a point in space and time, about 30-plus billion years ago, triggered the spherical-radiating chain reaction annihilations wave. Photons from annihilations radiate in all directions, and as they overlap and increase in intensity, they precipitate into subatomic particles that combine through several procedural steps to produce the matter we observe in the universe today. Precipitated particles formed at the speed of the photons (light) rapidly decelerate from the initial outward velocity by transfer of their linear momentums into angular momentums included in mixing and rotations.

Primordial matter particles are the positroniums, that is, mutually orbiting positrons and electrons. Paul Dirac and others first theorized their existence in the late 1920s (matter and antimatter). In a cloud chamber at California Institute of Technology in 1931, physicist Carl Anderson first observed positrons, the opposite electrical charge as an electron. Spacing of positroniums in primordial matter is such that its density is the same as observed in the universe today; thereby, the process is consistent with the first law of thermodynamics.

The matter we can observe today is limited to what exists inside the inner circle of figure II-7. We are located near the center and can only observe inside the inner circle, about halfway to the current position of the matter-transforming deflagration wave. Because the wave travel rate is almost the speed of light, the wave's actual location is about twice the distance of our see limit. All newly formed light-radiating objects are initially traveling at near the speed of light. They decelerate rapidly as linear momentums are transferred into angular momentums of rotation and orbiting objects and combinations.

Because we can only see the matter from which the light has reached our region of the universe, the observable universe is only about half the distance to the outer regions of the universe-building deflagration front's true and current location.

An abbreviated statement of the above verbal sketch is posted inside the front cover of *Model of the Universe*. The front cover is presented as follows.

Figure II-7. Model of the Universe with Laws of Physics.
(Same as reference 2 book cover)

Part III: Eureka! How, When, and Where the Universe Began

By definition, "void" means "having no contents." However, that is not the way it always was. The volume occupied by the origin void initially had the same matter content density as all primordial space and the subsequent universe. Voids are the exception.

I had been looking for such a void for over a decade. And in 2007, Lawrence Rudnick, professor at Minnesota University, stated he "discovered it by coincidence." The location is specifically where this new universe theory predicted ten years before it was discovered. However, Professor Rudnick found it to be about 7 billion light-years farther away than was originally estimated by this new concept. With recent Spitzer Telescope data, analyses shows this void and distance are proved compatible and feasible for the universe's start site. There appears to be several other void start sites, apparently starting from independent (?) spontaneous annihilations.

The Rudnick Eridanus Void has all the earmarks of the new universe theory's universe start site. The deflagration wave progressed through primordial matter positroniums, and then mass matter precipitated from the annihilation photons. The concept assumes primordial matter hexahedrons are approximately geometrically equilateral. The initial annihilation starting site resulting void would thereby be spherical within the limits of the uniformity of positronium spacing in the primordial matter.

Two pictures are available on the Internet of the 1907-discovered Eridanus Void. One shows the void as spherical; the other indicates a distorted oblate ball shape. Eventually, cosmological analyses will clarify this anomaly. The spherical void is used in this document, although an artist obviously produced the picture. The void's peripheral stars, in reality, would be much smaller. The universe's oldest stars are white dwarfs with their diameters about the size of Earth.

For perspective, one of the largest observable stars today, Betelgeuse, is a thousand times the diameter of our Sol, which is about 1,400 diameters of Earth. With the void diameter at 1 billion light-year, for the stars at the periphery to be of visible size in the void rendition, only a small segment of the void could be illustrated to scale on the same page, but it definitely shows the concept. Thanks to the void rendition producer(s).

Now, using Rudnick's void size and distance, we can estimate (and eventually define) the migration and deceleration rates of matter as it traversed the space between where it started and where it is now. Rudnick tentatively estimated the Eridanus Void to be about 10 billion light-years away and a billion light-year in diameter. The billion light-year-diameter void picture shows what it looked like 10 billion years ago if it were 10 or possibly 17 (figure II-3) billion light-years away.

However, the rendition picture shows what it may have looked like 17 billion years ago if it were 17 billion light-years away. Seventeen is the distance this new universe concept indicates by using figure II-3 as a nomographic. To simplify estimating, the distance of 10 billion light-years distance is tentatively used. Now, both distances are considerably farther than was assumed when this origin of the universe concept was developed and is farther away than is calculated using the obsolete Hubble number for calculating distance.

The deflagration wave, as described in some detail in *New Universe Theory with Laws of Physics*, also is briefly described in *Model of the Universe* and part II of this book. The wave is now believed to have started at what is the center of this void. After the initial annihilation, the deflagration wave propagated in all directions and transformed primordial matter into corporeal mass and energy. It continues to propagate in all directions at the speed of light. It also continues to precipitate mass matter at near the speed of light, but the precipitated mass matter particles immediately start accreting, coalescing, and mutually orbiting, which rapidly transfers linear momentums into angular momentums.

Thereby, all mass eventually loses most of its linear momentum. Rapid deceleration starts immediately when the second particle precipitates, and all subsequent objects and combinations continue to decelerate forever (Problem #10). The wave is now several (approximately 13.77) billion light-years beyond our part of the universe.

However, the Eridanus Void is observed as a billion light-year in diameter and calculated to be 10 billion light-years away, using the Hubble number[11] of seventy-three. The 10 billion light-year distance

---

[11]   It is believed that Lawrence Rudnick and his colleagues used the customary Hubble number of seventy-three or seventy-four to determine the distance. Hubble numbers were based on measured difference of velocities of different galactic clusters at different distances, which was misinterpreted as velocity rate versus distance. Now we know with proof as presented in Problem #8 of part I, galaxy clusters are actually decelerating. When the same galactic cluster's redshift is compared to the redshift of the same

to the Eridanus Void is more than three times the distance originally estimated for this origin of the universe concept. The light-radiating objects in the void's photograph near its periphery are about a half-billion light-years away from the void's central point, thus illustrating the rapid decelerations and slow matter migration speeds.

Also, the peripheral stars are estimated to be approximately 30-plus billion years old. In addition, the picture shows the void as it appeared 10 billion years ago. Our age is now approximately 30-plus billion years older than our start site, but we are now estimated to be only about 10 billion light-years from the void, proving our solar system, Milky Way galaxy, and local group clusters decelerated rapidly behind the wave, just like all precipitated physical/mass matter. The ongoing wave now continues progressing outward at light speed, and it is now far (many billion light-years) beyond us.

---

cluster fifty years later, the clusters are proved to be receding at a lower speed. (The chart in figure II-2 was used as a nomographic for estimates.) For our side of the void and velocity, using the Hubble number 117 should be more accurate. But when viewing objects in the other direction beyond the void, fifty-seven is expected to be more accurate. (Another value would apply between the void and us.) Rudnick estimated the distance to the void as 6 to 10 billion light-years. The gradient line for redshift versus distance is not linear and indicates the distance to the void from us is 10 to 17 billion light-years. For current simplification of explanations and estimating, the nominal distance, 10 billion light-years is the value used as distance to the void.

Figure III-1. Unknown artist's rendition for an optical view of the Eridanus Rudnick Void (image courtesy of NRAO).

The new universe theory (part II) explains how the deflagration wave process would leave such a void at its starting site. This site is farther distance away than presumed when this universe origin concept was conceived (references 1 and 2).

How the Void Formed

This new universe theory (Part II) explains that, before the universe existed, all space was filled with primordial matter, an endless hexahedron array of para-positroniums equally spaced at about a centimeter in each of all three dimensions. A positronium is a positron. An electron orbiting each other and together are often referred to as the simplest atom. The two are of equal mass and have equal but opposite electrical charges. And they are electrostatically and gravitationally attracted, but they are separated and stabilized by centrifugal forces from their mutual orbiting.

In primordial space, there are no extraneous radiations; thereby, positroniums remain stable until disturbed. (In the laboratory and everywhere within the universe, they are extremely unstable and quickly annihilate each other.) A single annihilation in primordial space initiated the wave of cascading annihilations, which continues to propagate in all directions. The wave (deflagration wave) includes the cascading annihilations and continues progressing through the endless hexahedron matrix of positroniums. Each annihilation results in a pair of gamma wave 511,000 electron volt photons, each radiating in opposite directions.

When sufficient numbers of forward-radiating photons adequately[12] overlap, they precipitate into elementary mass particles traveling at near the speed of light. Groups of particles gravitationally (also velocity-enhanced) coalesce, collide, spin, mutually revolve, and

---

[12]　It is not known how close to parallel traveling and what physical proximity photons need be to coalesce into mass. It is hoped the recently upgraded multibillion-dollar BEPC—the multination-funded Beijing Electron Positron Collider—in China may reveal new information on energy-to-mass conversions.

decelerate by giving up linear momentums to angular momentums while forming compound particles and objects. After many trillions of particle coalescences, clumping and via mutual self-gravity clumps become of adequate size to start and sustain nuclear fusions. Thus, first-generation stars were and are born.

Annihilation photons are made up of gamma rays that are the shortest of all electromagnetic wavelengths. Gamma wave photons that radiate in the reverse direction from the progressing deflagration wave front do not overlap adequately to coalesce and provide the higher intensity to precipitate into mass particles, resulting in true background radiation. These reverse radiating photons are energy packets that provide heat energy to dark matter. Dark matter radiates its latent heat in the spectrum that is known as CMB background radiation.

Microwave CMB radiation emanates from all objects that contain heat (thermally above absolute zero). This EMF energy is what the astronomy community has appropriately described as "background radiation." (Concepts that hypothesize that photons float around in the empty space is myth. In reality, empty space does not have heat capacitance and therefore has no microwave energy to radiate.) Gamma waves provide energy to dark matter for its heat to produce black body radiation, precisely as NASA's COBE spectrometer has measured and labeled as CMB.

The author speculates that most dark matter are stable multiple neutron nuclides, produced directly from photon transformations in the same manner that fusion processes in stars produce lithium, helium, and hydrogen ions, including deuterium and tritium.

The void photo shows the void region as it appeared when light from the photo object region emanated approximately10 billion years ago.

In real time, the void is different today, approximately 10 billion years since it was, as shown in the Eridanus Rudnick Void picture. The picture shows the visible mass matter had progressed from the center of the void to 500 million light-years in 20 billion years and in all directions as of approximately 10 billion years ago, as Rudnick calculated, based on the now-proven obsolete Hubble number of seventy-four. The picture, as seen today, is of the void as it was approximately 10 (or 17) billion years ago.

Based on the age of the universe as deduced from the (1) "timeline bar chart" (figure I-4) and (2) the velocity of objects decelerating behind the deflagration wave (figure II-2) and measured and reported in figures I-7 and I-8, and (3) the distance to the void based on the obsolete Hubble number of seventy-eight (figure II-3) and the void's diameter as reported and measured today as a billion light-year is now/currently 1.5 billion light-years in diameter.

Conclusion

Many astronomers will not want to recognize the void because it further adds to the big bang creditability problems, just like the other big bang problems (part I). However, they all support the deflagration wave concept. Actually, the deflagration wave concept (part II) was developed and conceived to be compatible with only factual observations and phenomena. One of the asymmetric observations (Problem #8) is another proof of major of asymmetry that is impossible with the unitary/point big bang source origin. The ten largest galactic clusters in the universe are located such that these larger older clusters suggest the direction to the origin void, whereas the smaller/younger galactic clusters are distributed in all directions.

The plotted data show where the origin void location should be found in the direction of right ascension 10:57 and declination +16:36. This is the direction where Professor Rudnick found what appears to be the large origin void in the Eridanus constellation.

The scenario for the universe void origin typically demonstrates the need for reviews, audits, and refinements of mankind's understanding with no myth and only the laws of physics consistent research studies. This document and scenario of the origin void demonstrates that all our (mankind's) cosmology "knowledge and understanding" needs to be audited, analyzed, and refined.

Questions

With part I undeniable plus the confirmation of no big bang and part II with the laws of physics consistent with the no myth concept, along with part III, the Rudnick Void discovery, we can now understand the universe's real development mechanics and dynamics. Also, we can now determine valid answers for many curious questions such as:

1. How far away is the deflagration front?
   - Traveling at near the speed of light, the annihilation front is now estimated to be a minimum of 40.5 billion light-years away from the origin void site in all directions. That is, our timeline plus our distance from the void plus the void radius. Beyond us, in our direction from the void, the wave is about 30 billion light-years from us. In the direction beyond the void from us, the wave is estimated to be about 45 billion light-years. The observable distances are about half the actual distances. The current actual distances are about twice the observable because of the time required for the light waves to reach us from the new stars following and behind the wave.

2. How far are we from the origin? What is our age?
   - The Rudnick team calculated our distance from the void to be between 6 to 10 billion light-years. They were not aware that the value they used to convert redshift measurements to distance was based on an obsolete Hubble straight-line value/number. The Hubble line is now shown to be curved, not straight, as assumed for Hubble numbers. The universe growth curve is now proven to not be a straight line, but it is now proven to be a curved gradient line based on directly measured galactic clusters recessional velocities. That curved

line is illustrated in figure II-3 and shows the distance to be between 10 and 17 billion light-years.

3. How large is the Rudnick Eridanus Void today (real time) by both distance and time?

- Assuming the void was a billion light-year in diameter, as of 10 billion years ago, at the decelerated migration speed of matter, it will continue to grow in all directions at 0.025 billion light-years per billion years. Therefore, the Rudnick Void diameter is now ([.050 x 10] + 1) = 1.50 billion light-years. This will be verifiable 10 billion years in the future and in another 10 billion years in the future. However, we and the solar system/Earth will no longer exist after another 4 billion years.

4. Where is the site/location from the Eridanus Void to where we (the local galactic cluster) began? Where was our mass and energy originally precipitated from primordial, as it applies to all mass and energy in the local group galactic cluster?

- Our timeline shows that our local galactic group (LGG) took approximately 30 billion years to reach the present maturity. The Rudnick Void photo presents the billion-light-year diameter for the void, as it was 10 billion years ago. Our distance to the origin void is tentatively accepted as 10 billion light-years. The stars at the void periphery traveled a half-billion light-years from the void center during their development time of 20 billion years. The oldest radiating stars are faded white dwarfs with minimum ages of 20 billion years, estimated after removing all slack time from their multigeneration star development time from our timeline (figure I-4). The rate of linear migration rate of stars at the void periphery are therefore 0.5 billion light-years/20 billion years = 0.025 billion light-years per billion years. At that rate, our LGG's region matter originally precipitated 0.75 billion light-years[13] closer to the void than we are now, or 9.25 billion light-years from the void's center and subsequently migrated to the current 10 billion light-years position (30 x 0.025 = 0.75 billion light-years).

---

[13]   Based on the timeline as shown in figure II-4, it required 30 billion years to produce mature white dwarfs to the dimness level, as Harvey Richer and his team observed. If all time increments as shown in the timeline chart were minimum and all slack times were removed, the most expedient required timeline could have been just 20 billion years. For the timeline for faded white dwarfs as seen in the Rudnick Void, twenty is assumed. The travel distance from the initial annihilation to the faded white dwarfs at the void periphery is measured as 0.5 billion light-years. This logically shows the migration rate to be (0.5 billion light-years /20 billion years = 0.025 billion light-years/billion years). Because we are 10 billion light-years from the void, at that same migration rate, we would have taken 400 billion years (10 billion light-years / 0.025 billion light-years) time to reach this position. Also, using our applicable timeline of 30 billion years, (30 x 0.025), our Local Group Cluster mass matter would have started 0.75 billion light-years back toward the void from our current position. That is the distance we have migrated in our timeline development of 30 billion years. Using the 10 billion light-years distance to the void, the start site would be 9.25 billion light-years back from the void. [Note: These seem like excessively small and large numbers, but it is to be remembered that the numbers are billions of years and billions of light-years. And we are a relatively insignificant component of the universe.]

Epilogues I and II

The title for this book, *Proof*, was recommended by Ronald Mobley, a respected, successful, and intelligent entrepreneur friend formerly from Kansas City, Missouri, and subsequently Bremerton, Washington, and now Surprise, Arizona. He thinks this name will trigger curiosity among intelligent people. On the other hand, another colleague, Robert E. Farrell, who has a PhD in engineering, also ardently supports this concept and said the name may offend dogma-inflicted individuals. They may refuse to read it. I choose to think and hope that intelligence will win out over prejudice and dogmatism.

Dr. Farrell convinced me that the title *Problems* would hopefully stimulate thinking rather than resistance and dogma defense of the big bang concept.

Epilogue I: My First Big Bang Exposure

In 1932, I was five years old when my mother allowed me to get out of bed in the wee hours to listen to a special shortwave broadcast from England if I agreed to be quiet. (It was no easy chore!) Dad was at work on his local newspaper linotype/printer job from 5:00 p.m. to 2:00 a.m. Dad had set up his shortwave radio in the kitchen so my older brother Ernie and three of his buddies could listen to a special program to be broadcast by a professor of physical sciences, Sir Fred Hoyle. Ernie, then thirteen, and his buddies[14] were going to listen to the broadcast. I was instructed to lie quietly on the kitchen floor in my pajamas. Hoyle apparently made regular broadcasts about the universe in the early 1930s, which we received after midnight in the wee hours in Enid, Oklahoma. Hoyle was an impressive lecturer, and during the broadcast, he stated, "The big bang is bunk."

Before that broadcast, some astronomers, including Edwin Hubble, claimed the universe all erupted from a single point in space to become the universe. Fred Hoyle derogatorily called it "just a big bang," and the name stuck as the name of that concept. After the broadcast, my brother and his buddies agreed the concept was bunk. My mother was present and admonished them for using such bad language, but after much discussion, she relinquished as Ernie and his buddies convinced my mother it was okay to say "bunk" when talking about the big bang.

Sorry, Sir Fred Hoyle didn't live to find out he was right after all. I think he knew all along. He just didn't have any other explanation at that time.

---

[14] Jack Coffin, Jack Helton, and Everett Gunning, I am saddened that my brother and his buddies are not still around to read the proof presented in my book III. They provid much of my ongoing motivation.

96

Epilogue II: Philosophy of Mankind's Understanding

Valid understanding is only via scientific thinking. Scientific thinking uses logic with only proven facts and verified and repeatable phenomena. Verified phenomena is proven many times but never disproved becomes laws of physics. To believe is to accept without proof. Myths cannot be proved or disproved.

Statements by Robert E. Farrell and Bobby McGehee

Getting this proven real universe origin accepted is an uphill battle. Our government (NASA) has spent a lot of money justifying and promoting the big bang. It does not matter what is right but rather where the funding is. Cosmology researchers don't want to expose their prior belief, and they need funding.

Some still subscribe to the flat earth concept! Most of us say, "How ridiculous!" It may take several lifetimes for this proved universe origin realism to prevail.

Bibliography and References (Group 1)

Light travels faster than sound. That is why one may appear bright until they speak.

—Unknown origin of quote

Our apologies to those who contributed yet were missed in our bibliography and listing of references.

1. Bobby McGehee, *"New Universe Theory with Laws of Physics" AuthorHouse,* Bloomington, Indiana, 2005.
2. Bobby McGehee, *"Model of the Universe" AuthorHouse,* Bloomington, Indiana, 2010.
3. James C. Baker, ME, a rocket scientist for the Boeing Company and NASA, *Stated in email correspondence with author* said, "It is more logical to assume that everything developed from space and matter that already existed than to assume everything came from nothing out of a single point." 2009
4. Kenneth C. Freeman, "The Hunt for Dark Matter in Galaxies." *Science* 302 (5652): 1902–1903. *Paper presented at AAAS Meeting,* December 12, 2003
5. Steven Weinberg, *"The First Three Minutes"* New York: Basic Books, 1977.
6. Alan H Guth, PhD. *"The Inflationary Universe".* Cambridge Mass.: Perseus Books, 1997.
7. Harvey Richer, et al. "Study Indicates Some Stars Are Older Than The Universe!" *Astronomy,* 1997.
8. Dan Brasoveanu, PhD. *"Modern Mythology and Science",* New York: iUniverse, Inc., 2008.

Bibliography and References (Group 2)

1. Bobby McGehee, *"New Universe Theory with Laws of Physics"* Bloomington, Indiana,: AuthorHouse, 2005.

2. Bobby McGehee, "Universe Milestone Bar Chart." published by McGehee, *Model of the Universe with Laws of Physics* (2010).

3. John Huchra home page. "Hubble number listings 1928–2007." http://www.cfa.harvard.edu/~huchra.*(Website may have been replaced at Harvard;* John Huchra deceased in 2009)

4. Bobby McGehee First estimate for our distance to the center of the universe, Figure 7.7 of *"New Universe Theory with Laws of Physics"*: 2005.

5. Michael Richmond's Supernova Page. "SNe after 1988 to 4.16.2002." www.tass-survey.org/richmond/sne/sne.org.*(Website may not be available)*

6. Reiss, Adam, G., Strolger, Louis-Gregory, Tonry, John, Casertano, Stefano, Ferguson, Henry C., Mobasher, Bahram, Challis, Peter, Filippenko, Alexei V., Jha, Saurabh, Li, Weidong, Chomock, Ryan, Kirshner, Robert P., Leibundgut, Bruno, Dickerson, Mark, Livio, Mario, Giavalisco, Mauro, Steidel, Charles C., Benitez, Narciso, and Tsvetanov, Zlatan. "Type Ia Supernova Discoveries at z = 1 From the Hubble Space Telescope: Evidence for Past ..." arXiv:astro-ph/0402512v2 (2004).

5. Reiss, Adam G., Strolger, Louis-Gregory, Casertano, Stefano, Ferguson, Henry C., Mobasher, Bahram, Gold, Ben, Challis, Peter J., Filippenko, Alexei V., Jha, Saurabh, Li, Weidong, Tonry, John, Foley, Ryan, Kirshner, Robert P., Dickinson, Mark, MacDonald, Emily, Eisenstein, Daniel, Livio, Mario, Younger, Josh, Xu, Chun, Dahlen, Tomas, and Stern, Daniel. "New Hubble

Space Telescope Discoveries of Type Ia Supernovae at Z >1 ..."
arXiv:astro-ph/0611572v2. (2007).

6. The Bootes Void. www.acceleratingfuture.com/michael/ blog/?p=69. *(Reference data available via Bing search "Bootes Void")*

7. Rudnick, Lawrence; Brown; Shea; and Williams, Liliya R. "Extragalactic ... and the WMAP Cold Spot." arXiv 0704.0908v2 [astro-ph]. (August 3, 2007).

8. British UK team: Jack Milton (UK), Kate (UK), Issak (Clavendish Laboratory U of Cambridge), Robert Priddey (Imperial College in London), and Richard McMahon (U of Cambridge), reports in April 2003 that the most distant quasar highest redshift (Z = 6.43) Farthest Quasar at 87 percent of c.

9. Frebel, Anna. *Quasars.*

10. Whitaker, Stephen. *Introduction to Fluid Mechanics.* Malabar, Fl.: Krieger Publishing Co., 1968.

11. Marks, Lionel S. *Mechanical Engineers Handbook.* McGraw-Hill Book Company, 1941.

12. Catalog of Abell Clusters by George Abell (Cal Tech), Harold Corwin (Cal Tech, formerly University of Edinburg), and Ronald P. Olowin (University of Oklahoma).

13. Baum, Edward M., Knox, Harold D., and Miller, Thomas R.; "Chart of the Nuclides, 16th edition; "ChartOfTheNuclides.com"; also see "Chart of Nuclides"

14. Weinberg, Steven. *The First Three Minutes.* New York: BasicBooks, 1977.

15. Particle Data Group of Lawrence Berkeley National Laboratory. "History of the Universe Poster." http://particleadventure.org.

16. Bobby McGehee, "Deflagration Wave, Figure 6.14, from *New Universe Theory with Laws of Physics*, 2005

17. Beijing Electron Positron Collider (BEPC). www.ihep.ac.cn/ English/E-BEPC/index.htm.
18. Cosmic Background Explorer (COBE). www.britannica.com/eb/ topic-139203/Cosmic-Background-Explorer.
19. Fowler, Michael. "Black Body Radiation." http://galileo.phys. virginia.edu/classes/252/black_body_radiation.html.
20. "Planck's Law of Black Body Radiation. http://en.wikipedia.org/ wiki/Planck%27s_law_of_black-body_radiation.
21. "WMAP Glossary of Technical Terms." http://map.gsfc.nasa. gov/m_help/h_glossary.html.
22. Richer, Harvey. "M4 White Dwarf Stars Cool." http://antwrp. gsfc.nasa.gov/apod/ap971102.html and http://antwrp.gsfc.nasa. gov/apod/ap950910.html.
23. SETI is the name used for several projects searching out extraterrestrial intelligence. The first private SETI organization was an offshoot from the Planetary Society, which Carl Sagan, Bruce Murray, and Louis Friedman founded in 1980. (I was a charter member.) Space Science Laboratory (SSL), which was initiated in 1958 at UC at Berkeley, now coordinates SETI@home efforts.
24. Richmond, Michael. International Supernovae Network. www. supernovae.net/isn.htm.
25. *Science*, a publication of American Association for Advancement of Science (AAAS), published a solicitation for Internet users to help classify galaxies catalogued from the Sloan Digital Sky Survey (SDSS) telescope in Sunspot, New Mexico. The web site is Galaxy Zoo, www.galaxyzoo.org.
26. List of Abell clusters (includes all ACO clusters). http:// en.wikipedia.org/wiki/Abell_catalogue
27. www.chemicalelements.com

28. Holden, Norman E. "History of the Origin of the Chemical Elements …" 2004.

29. Particle Data Group of Lawrence Berkeley National Laboratory (LBNL). "History of the Universe Poster." particleadventure.org/frameless/history-universe.html.

30. Sorbel, Dava. *Galileo's Daughter*. New York: Penguin Books, 1999.

31. Tirion, Wil. *Sky Atlas 2000.0*. Cambridge Mass.: Sky Publishing Corporation, 1999.

32. Sagan, Carl. *Cosmos*. New York: Random House Inc., 1983.

33. Strasbourg Astronomical Data Center (CDS). The Smithsonian/NASA Astrophysics Data System. 1972.

34. Coe, Steven. "Discover Galaxy Groups and Clusters." *Astronomy* (March 2009).

35. Struble and Rood. "1999 ApJS.125 … 35S." *NASA /IPAC Extragalactic Database (NED)*.

36. Freeman, Ken C. "The Hunt for Dark Matter in Galaxies." *Science* (December 12, 2003).

37. Sternglass, Ernest J. *Before the Big Bang*. New York: Four Walls Eight Windows, 1997.

38. Paul Adrien Maurice Dirac was born on August 8, 1902. Dirac's work has been concerned with the mathematical and theoretical aspects of quantum mechanics. He wrote a series of papers on the subject, published mainly in the Proceedings of the Royal Society, leading up to his relativistic theory of the electron (1928) and theory of holes (1930). This latter theory required the existence of a positive particle having the same mass and charge as the known (negative) electron. This, the positron, was discovered experimentally at a later date (1932) by C. D. Anderson, while its

existence was likewise proved by Blackett and Occhialini (1933) in the phenomena of "annihilation."

39. NASA / IPAC (National Extragalactic Database or NED) includes the CDS that currently has on file about 8,232 catalogues of stars, galaxies, and extragalactic objects, including the Abell (ACO). The Strasbourg Astronomical Data Center (CDS) [36] was created in 1972 to collect data concerning astronomical objects, making it available on electronic form.

40. Scott, Douglas..

Bibliography and References (Group 3)

1. Lal, Ashwini Kumar, PhD. "A Critical Review of the Big Bang." *Journal of Cosmology* 6 (2010): 1533–1547.
2. Jastrow, Robert, PhD. *Red Giants and White Dwarfs: The Evolution of Stars, Planets, and Life.* Harper & Row, 1967.
3. Farrell, Robert E., PhD. *Alien Log II.* Sun City West, Ariz.: R. E. Farrellbooks LLC, 2011.
4. Lerner, Eric J. *The Big Bang Never Happened.* New York: New York Times Books/ Random House and at Random House or Canada Limited, 1991.

Theory Developer and Author

Engineering physicist Bobby L. McGehee, the theory author of this text, was born in Enid, Oklahoma, on September 19, 1927. He served in the USMS and USAF. He studied at Oklahoma State University, qualifying for degrees in engineering physics, physics, and education. He taught basic physics as a student instructor. During his thirty-two year career in the aircraft industry (Beech and Boeing), he was project engineer for development, design, construction, and operation of propulsion and aerodynamics test facilities. He was honored with the following awards: Engineer of the Month and Engineer of the Quarter out of a staff of 2,500. Also, he is an associate fellow of the American Institute of Aeronautics and Astronautics and served on the board of directors.

After retiring, he returned to college to study astronomy and geology. He is a recent member of the Astronomical Society of the Pacific, American Association of Physics Teachers, and American Association for the Advancement of Science.

Bobby McGehee can be reached at BNMcGehee1@msn.com

***

In the 1950s, McGehee refused to accept what he called "the impossible big bang" and discontinued studying astronomy as a science. After a successful engineering physics career, he returned to college in Seattle to study geology, chemistry, and astronomy. His post-retirement studies and research led to his development of a scientific origin of the universe concept, which is consistent with the proven laws of physics. Two books authored by Bobby McGehee, *New Universe Theory with Laws of Physics* and *Model of the Universe*, further describe this scientific concept.

McGehee has presented these books and theory via booths at national AAPT and ASP meetings and given lectures at local science clubs and community colleges. At the West Valley Engineers presentation, colleague Robert E. Farrell immediately accepted and endorsed this new universe theory. Since the publication of *Model of the Universe*, Robert E. Farrell has contributed to ongoing studies.

McGehee is a current or past member of local science organizations, and national AAPT (Amrican Association of Physics Teachers), ASP (Astronomical Society of the Pacific), and AAAS (American Association for the Advancement of Science), and he is a past member of the board of directors of AIAA/PNW (American Institute for Aeronautics and Astronautics / Pacific Northwest).

Bobby McGehee's books are available at bookstores. Book copies are also available in hardback, paperback, and e-book at AuthorHouse. com/bookstore.

Closing Comments:

Universe's True History based on Proven Event Chronology:

This is known only by tangible evidence, direct observation, and scientific logic. (Scientific logic includes the use of only proven facts and phenomena, the laws of physics and no myth. The origin from primordial matter started more than 30 billion years ago. When and how? The universe began 25–35 billion years ago. What was the cause? When and where were the initial annihilation? The site was Eridanus Void. First-generation stars, fusion process, and supernovae converted hydrogen. For second-generation stars; fusion and supernova produces light elements. For third-generation stars, fusion produces heavier elements. Supernovae produces heaviest elements.

Fourth-generation stars and planets precipitate/form, for example, Sol and Earth, 4.5 billion years ago. One planet includes water and produces atmosphere with 21 percent oxygen. A large planetoid (Mars-sized) collides with Earth and rebounds to form moon and tilt earth, which causes seasons. Replicating compounds have been found which are 3.5 billion years old (stromatolites). Biological compounds occurred 3-plus billion years ago. Replicating and mobile biological forms appear 2 billion years ago. Mobile life forms started evolving. A large meteorite strikes and destabilizes core. It fractures shell and initiates continental drifting 0.825 billion years ago. Life evolves into cold-blooded, large, amphibian reptiles (dinosaurs) and warm-blooded, shrew-sized mammals.

Meteorite strikes Earth at north shore of Chicxulub Peninsula 80 million years ago, causing three-plus years of darkness. Climate change included ice age and minimal vegetation growth. Large vegetarian and carnivore eaters starved. Small mammals (shrew-sized) survive, evolve, and proliferate. Humanlike beings first appeared over 125,000 years ago. Civilizations started. Evidence implies civilizations existed at least forty thousand years ago. Leaders controlled populaces with myths and such civilizations intermittently ceased and restarted due to environmental changes. At least one civilization began over twelve thousand years ago. A buried pyramid in China was discovered after 2000 AD. Current civilizations existed eight thousand years ago. Sunken city discovered at the mouth of the Nile River in 2010. Knowledge of how the universe began has now been deducted and is now proven in this 2015 publication. No big bang or other human invented myths are included.

Note

Lists of references that support each milestone and milestone event time refinements would be appropriate here, but are left for students and others.

Origin Site of the Universe based on facts, logic, 2007 discovery: Rudnick Eridanus Void, Drawing Rendition Credit NASA/NRAO, Billion-Light-Year-Diameter Void at RA10:57, Dec+16:36, located approximately 10 to 17 billion light-years distance.

Discoverer, Professor Rudnick, Minnesota University, 2007. Revealed 2009, Confirmation 2014. If this void is verified, it alone will be another undeniable proof the big bang could not be the universe origin. The big bang concept is impossible with this void. The void is believed to be the universe origin, and its existence is confirmation of the "laws of physics" universe's origin concept presented as Part III of this book. The universe started growing from what is now the center of this void, and it started approximately 30 billion years ago. The universe is now approximately 60 billion light-years in diameter.

The stars at the edge of this huge void are the oldest in the universe. In a to-scale rendition, these oldest stars could not be seen on the same page with this void drawing. If stars are drawn as one millimeter, the billion light-year void dark area would need to be drawn on a page 1,021 miles square! Obviously, only a very small segment of the Rudnick Void could be drawn to scale on the same page with the universe's oldest stars. This void is huge! It would take ten thousand Milky Way galaxies to bridge such a gap.

The universe's age/milestone bar chart (figure I-4) substantiates the universe's age at approximately 30 billion years. The universe began at tis void's center approximately 30-plus billion years ago, and the universe's diameter is now approximately 60 billion light-years.

Some of the oldest stars in the universe are as found in globular cluster M4 and M2. They are faded white dwarfs, which are burned-out fourth-generation stars. These were stars like our sun, which, upon depleting enough of their fuel and the concurrent reduction of self gravity, they erupted as a planetary nebula. The remaining core settled into an inert white-hot, white light-radiating star remnant about the size of the Planet Earth (between approximately six thousand and approximately ten thousand miles in diameter). It will take forever for them to radiate all of their heat.

I do not know how the NRAO determined/measured the diameter of this void. If their measurement was determined using radio or other electromagnetic waves, the measurement waves would have left that site 10 to 17 billion years ago since it was determined to be 10 to 17 billion light-years away. Our solar system is proven to be almost 30 billion years of age. Therefore, the void diameter would now, since the electromagnetic waves left the site, have grown to a diameter of approximately 1.5 billion light-years..

The distance was likely determined using obsolete Hubble numbers. Hubble numbers are based on the now obsolete big bang concept, disproven by the existence of this void as well as several laws of physics facts. Therefore, it would actually be farther away, estimated at 17 billion light-years away, based on realistic/curved Hubble line as shown on Figure II-3.

Prior to June 2014, Lawrence Rudnick wrote in an email he was skeptical about the void being real. However, in a more recent email received on June 19, 2014, he wrote, "As of this week, there may be confirmation!"

Update for JWST

The JWST new technology space telescope will be launched in 2018, and become operational 2019. Unlike the Hubble space telescope, this telescope will orbit the sun in the shadow of the earth. In addition to being always in the earth's shadow it will be cryogenically cooled and thereby have sensitivity to measure far red optical line spectra data not otherwise possible. The concept and observation site allows acquisition of six digit red shift 'z' data. JWST will have ~25 times more mirror area than the successful Spitzer space telescope in the same environment. Spitzer telescope depleted its cryogen supply in 2012, however, JWST includes refueling capability. The project science team engineering and managing the JWST into existence includes 20 top cosmology specialists from several countries. I was invited to attend and participate in planning meetings for defining JWST projects.

These proposed Study/projects will prove the 1920's big bang universe origin concept obsolete and will verify a deflagration wave that precipitates mass matter from primordial matter. Unlike the Big Bang, this concept is Laws of Physics consistent, without myth.

The following projects are recommended as high priority because the results will influence all cosmology studies, past and future.

Proposed projects for Early JWST study:

1. *Demonstrate universe matter is decelerating,* contrary to 1920's Hubble assumed acceleration. *Obtain 2019 time lapse velocity change measurements of the same 30 galactic*

clusters as measured with the Spitzer space telescope in 2009 and by Palomar in 1999 and 1959. This will prove BB did not happen, as thought in the 20th century. 2019 JWST Z data compared to 2009 Spitzer Z data of the same 30 ACO galactic clusters*, will more precisely define deceleration rates. With data acquired in the opposite direction to the start void, Spitzer & JWST data can prove very high deceleration starts immediately behind the wave where precipitated mass objects' enhanced gravity causes objects mutual orbiting converting linear momentum to angular. 80 to 90 % of initial velocity is converted within the first .5 billion years. Further confirming the deflagration wave concept.

* *Listed in Ref 3, pgs 33 & 35.*

2. **Determine our absolute velocity through space.** *Measure velocity of matter on both the near and far sides of Rudnick void,* presumed to be our universe's deflagration wave origin site. It is theorized mass matter initially precipitated and coalesced near c velocity behind an outward progressing deflagration wave as it transcends primordial matter. It is estimated to be about .25 c at our distance from the initial site. This measurement should confirm this void as the origin.

3. **Establish and define a valid new "Hubble line".** Revise and correct our 'Hubble line' yard stick for determining distances to various astronomical objects. Presumed by Hubble to be a straight line to infinity. A draft/estimated line (shown on Fig II-3 of ref) is an approximate distance vs velocity line for objects in line with our origin-radiant line. The new line should be a curved gradient line starting from the universe's

origin site* extending to mass' current precipitation velocity. That location is mass matter's initial location behind the outward progressing deflagration wave. On a graph of velocity vs distance. All distance measurements between us and an object need be corrected for the offset between our origin radiant line to the object. (line flagged by arrow '#2' in Figure II-3 of Ref 3).

4. **Measure** *universe's observable and physical sizes and radius at various directions.* Measure our separation rate and direction from the origin site. By assuming spherical propagation of the deflagration wave precipitated matter we can Calibrate the universe's outward deceleration. (From our off-center position the universe periphery should be an oblate spheroid.) BB thinking had everything at the center, but we now know we are about 10 to 17 billion light years beyond our origin site which is near the origin void. The universe's 'diameter' is estimated to be 8 times larger than by BB thinking.

Note: Matter development time to produce the local group cluster is ~30 Billion years (Ref 1, fig I-4), which started when 'our stuff' precipitated from the wave. We are apparently ~10, (maybe 17+) Billion Light Years distance from the origin void. The universe is growing radially at the speed of the deflagration wave (light); it is approximately 30 + 10 Bly radius. (We can observe ["see" light radiating from objects] only within the observable limit). The universe periphery is (25 + 10) Bly away from us in our propagating direction, and [25 + (2x10)] in the opposite. With adequate optical instruments, the Observable limit is ~15 Bly this side of the void, and ~25 in the opposite direction.

5. **Define a family of velocity vs distance profile lines for** mass matter as it precipitated, decelerated and propagated following the wave. The Rudnick void was just discovered in the first decade of millennium 2000 and appears to be the starting site of the deflagration wave. It is huge at one Bly light year diameter (.5 Bly radius). It is estimated by NRAO to be ~10 Bly distance. If this distance measurement is valid, Its diameter has grown to ~1.25 Bly while the light from it was traveling to us. The initial mass matter is at the periphery of the void, and the universe's age is known via knowledge and understanding of sequential mass transformation processes, the universe is over 30 Billion years of age (See Fig I-4. pg 26 of ref 1)

The initial mass' matter from the start of the deflagration wave is still, and always will be, at the periphery of the void. We know our matter has decelerated from near c to its current velocity very rapidly with respect to distance, (from .999+c to <.3c in only ~1 to 300,000,000 years). obviously the ongoing separation velocity is relatively low. Transfer of linear momentum to angular is the only velocity changing force. Initial / original matter is progressing outward at a continually slowing speed. At this point most matter have been accreted into angular momentums (vortexes), and the deceleration almost, but will never cease. Yet the deflagration wave powered by annihilations continues outward progression at its velocity near c.

6. *Determine and define blended regions of overlapping multi-universes. 'Multiverses' could and would exist if there are more than one deflagration wave start site (voids) and there appears to be at least two, maybe three.* Measure

velocity of matter in the vicinity of Abell void (both nearer and farther) to determine if we have multiple universes that are mixing/blending and resulting in various velocities of decelerating matter in overlapping regions behind potential multiple interacting deflagration waves.

7. **Via the internet, it was revealed that a large ~2.5+ Bly void has been detected. If the discovery is valid, it is more likely the universe origin as the size would better fit the logic for time requirements for 'cascading' multi generation star development. We do know the mere existence of voids and the big bang concept are not compatible and could not coexist. The void origin site does not require hypothetical myth and the BB does, plus the BB is not consistent with proven laws of physics. Of course we will still ask what was the beginning of the primordial?**
   Defining true geometry, mechanics, and dynamics of the universe is an enormous task but JWST can give humanity a valid start. I hope I will be around for the next decade to see JWST reality!

Supporting books by Bobby McGehee. All are available from amazon.com, most book stores, in paperback, hardback, and e-book. <www.authorhouse.com/ bookstore>

1. **New Universe theory** (**N. U. T.**) Written during the development of the new theory of the universe. Theory supersedes the 1920's big bang idea in favor of Laws of Physics and without myth. Origin of primordial matter and primordial space will always remain as the ultimate question. [C 2000]

    Softcover: 1418494291;   Hardcover: 1418494305;   Ebook: 1418494313

2. **Model of the Universe**. With Laws of Physics Conveys geometry of the universe. Presents proof (40 year time span data) the universe is not expanding with acceleration. Uses Palomar data. [C 2005]

    Softcover: 9781449067922;   Hardcover: 9781419067939;
    Ebook: 9781449067946

3. **Big Bang Problems** / **How the Universe Began**. Presents several big bang credibility problems and proves the universe's growth is not accelerating. It is decelerating, proven over a 50 year time span using 1959 Palomar and 2009 (Spitzer) data. Book also presents recent discovery about the universe origin site(s?) [C 2014]

    Hardcover: 9781496946416;   Softcover: 9781496946461;
    Ebook: 9781496946423

The JWST could only come into existence thanks to the dedicated efforts and contributions by hundreds of capable motivated people.

As a tribute to all contributing personnel, it is proposed that a plaque be placed on the JWST before launch. The plaque should have the names of the Science team who made the Telescope possible, as representatives for all contributors. This plaque will exist beyond all humanity, until Sol ages to its inflation, expected to occur some 3+ billion years hence.

Science team* includes at least 20 cosmology specialists. *THANKS*

JWST SCIENCE TEAM

John C. Mather *	NASA/GSFC;	<u>Chair</u> & <u>Senior Project Scientist</u>
Mark Clampin	NASA/GSFC;	Observatory Project Scientist
Rene Doyon	U of Montreal;	CSA Project Scientist
Pierre Ferruit	ESA;	ESA Project Scientist
Kathy Flanagan	*STScI;	Acting Director, STSci
Marijn Franx	Leiden University;	NIR Science Rep
Jonathan P. Gardner	NASA/GSFC;	Deputy Sr. Project Scientist
Mathew A. Greenhouse	NASA/GSFC;	ISIM Project Scientist
Heidi B. Hammel	AURA Space Science Inst;	interdiscip. Scientist
Simon J. Lilly	Swiss Fed inst. of Tech.;	Interdiscip.Scientist
Jonathan I. Lunine	Cornell University;	Interdiscip. Scientist
Mark J. McCaughrean	ESA;	Interdiscip. Scientist
Matt Mountain	AURA/STScI Director	Telescope Scientist
Malcolm Niedner	NASA/GSFC;	Dep. Senior Project Scientist/Tech
George P. Reike	University of AZ NIRC	Principal Investigator
George Sonneborn	NASA/GFSC	Operations Project Scientist
Massimo Stiavelli	STScI	Interdiscip. Scientist S&CC Lead
Rogier A. Windhorst	Az. State University;	Interdiscip. Scientist
Chris J. Willott	Herzberg Inst. of Astrophysics;	NIRISS Science Ld
Gillian S. Wright	UK Astronomy Tech. Ctr.	MIRI; EU Science Ld

List to be edited/revised

Milestone Contributors 20th & 21st Century Cosmology Knowledge

V M Slipher, 1912 *	Galaxy red shift increases w/ distance
Milton Humason, 1918 *	First to Document velocity vs distance
Edwin Hubble, 1923	Hubble#, red shift vs distance O.K to.5c
Fred Hoyle, 1931	Named the concept; 'Big Bang is Bunk'
Albert Einstein, 1915 *	Mass to/from energy, relativity effects
Paul Dirac, 1930 [1]*	Discovered Positrons and Positroniums
George Abell, 1947	Palomar All Sky Survey Catalog #1
Harold Corwin, 1959	Palomar All Sky Survey Catalog # 2, 3
Ronald Olowin, 1960	Palomar All Sky Survey Catalog #'s 2 +
Harvey Richer, 1995 [2]	Faded White Dwarfs are older than BB
Kenneth Freeman, 1995 [3]	Measured Dark Matter in galaxies, 87%
Herbert Rood, 1999 [4]*	R & S Cataloged 40 year time lapse 'Z's
Mitchell Struble, 1999 [4]*	R & S Cataloged 40 year time lapse 'Z's
Spitzer Team, 2009 [5]*	Cataloged 6 digit cluster redshift Z data
Lawrence Rudnick, 2009 [6]*	Discovered Eridanus Void, 1 Bly dia.
Bobby McGehee, 2010 [7,3]*	Galactic clusters are decelerating,50 yrs
JWST Science Team, 2020	JWST Telescope...Universe geometry

* *Discoveries that contribute to proof the Big Bang never happened.*

[1] *Positrons are anti-matter electrons Mutually orbiting are positroniums.*

[2] *Defines real-time increments required for development of universe age, age of White dwarf exceeds the maximum for BB universe's age.*

[3] *Proved dark matter is real. BLMc Speculates multiple neutron nuclides bound together by bosons. Gaseous cloud collections by gravity.*

[4] *Time lapse re-measure of red shift provides data for universe dynamics.*

[5] *Time lapse re-measure of red shift with 6 digit data proves deceleration.*

[6] *Voids are impossible with point origin.BB and Voids are not compatible.*

[7] *50 yrs Deceleration directly disproves the BB point source concept.*

Bobby McGehee

JWST Vital Facts:

Launch Date	October 2018
Launch Vehicle	Ariane 5 DCA
Mission Duration	5 - 10 years
Total payload mass	~6200 kg
Dia. of Primary Mirror	~6.5 m (21.3 ft)
Clear Aperture	25 m2
Primary mirror material	beryllium, gold plate
Mass of primary mirror	705 kg
Mass; single mirror segment	20.1 kg
Focal length	131.4 meters
Number of segments	18 hexagons
Optical resolution	~.1 arc seconds
Wavelength coverage	0.6 - 28.5 micro
Sun Shield size	21.2 x 14.2 m
Orbits Sun beyond earth	1.5 million Km
Orbit in earth's shadow	LaGrange Point # L 2
Orbit temperature	<50 deg K,-370 F

JWST Specialized Instruments:

Near Infrared Camera (NIRI)
Near Infrared Spectrograph (NIRISS)
Mid Infrared Instrument (MIRI)
Fine Guidance Sensors
Slitless Spectrograph (FGS/NIRISS)

Desired objective: Confirm all universe dynamics and mechanics are consistent with Laws of Physics

Image by courtesy NASA/National Radio Astronomical Observatory Image. (Printout from the internet included this address: <03 08 07_hubble_collisions_04 http://www.bing.com/images/ search?q=void+Rudnick&FORM=1G ... 10/18/2013> (Recently the photo could not be found either by myself or Professor Lawrence Rudnick)

Printed in the United States
By Bookmasters